Red Coats and Wild Birds

FLOWS, MIGRATIONS, AND EXCHANGES

*Mart A. Stewart and Harriet Ritvo, editors*

The Flows, Migrations, and Exchanges series publishes new works of environmental history that explore the cross-border movements of organisms and materials that have shaped the modern world, as well as the varied human attempts to understand, regulate, and manage these movements.

# Red Coats and Wild Birds
*How Military Ornithologists and Migrant Birds Shaped Empire*

Kirsten A. Greer

The University of North Carolina Press  CHAPEL HILL

*Publication of this book was supported in part by a generous gift from Kim and Phil Phillips.*

© 2020 The University of North Carolina Press
All rights reserved
Set in Merope Basic by Westchester Publishing Services
Manufactured in the United States of America

The University of North Carolina Press has been a member of the Green Press Initiative since 2003.

Library of Congress Cataloging-in-Publication Data
Names: Greer, Kirsten A., author.
Title: Red coats and wild birds : how military ornithologists and migrant birds shaped empire / Kirsten A. Greer.
Other titles: Flows, migrations, and exchanges.
Description: Chapel Hill : University of North Carolina Press, [2020] | Series: Flows, migrations, and exchanges | Includes bibliographical references and index. | Summary: "During the nineteenth century, Britain maintained a complex network of garrisons to manage its global empire. While these bases helped the British project power and secure trade routes, they served more than just a strategic purpose. During their tours abroad, many British officers engaged in formal and informal scientific research. In this ambitious history of ornithology and empire, Kirsten A. Greer tracks British officers as they moved around the world, just as migratory birds traversed borders from season to season"— Provided by publisher.
Identifiers: LCCN 2019019412 | ISBN 9781469649825 (cloth : alk. paper) | ISBN 9781469649832 (pbk : alk. paper) | ISBN 9781469649849 (ebook)
Subjects: LCSH: Blakiston, Thomas Wright, 1832–1891. | Ornithologists—Great Britain—History—19th century—Biography. | Migratory birds—Social aspects—Mediterranean Region. | Great Britain—Armed Forces—Officers—History—19th century—Biography. | Great Britain—Colonies—Geography. | Great Britain—Colonies—History—19th century. | Mediterranean Region—History—19th century.
Classification: LCC DA68 .G825 2020 | DDC 355.0092/241—dc23
LC record available at https://lccn.loc.gov/2019019412

Cover illustration: Cornelius Krieghoff, *An Officer's Room in Montreal* (oil on canvas), 1846. Used with permission of the Royal Ontario Museum © ROM.

*I dedicate this book to my daughter, Annsophie.*

# Contents

Acknowledgments xi

Introduction 1

**CHAPTER ONE**
Red Coats and Wild Birds across the British Empire 10

**CHAPTER TWO**
Thomas Wright Blakiston 23
*Crimean Scientific War Hero*

**CHAPTER THREE**
Andrew Leith Adams 41
*Mediterranean Semitropicality*

**CHAPTER FOUR**
Leonard Howard Lloyd Irby 63
*British Military Ornithology on the "Rock"*

**CHAPTER FIVE**
Philip Savile Grey Reid 81
*Red Coats and Wild Birds on the Home Front*

**CHAPTER SIX**
Military Ornithology in Place 97
*Territoriality, Situated Knowledges, and Heterogeneities*

Afterword 104
*Avian Colonial Afterlives*

Notes 109
Bibliography 141
Index 167

*Figures and Maps*

FIGURES

1. Cornelius Kreighoff, *An Officer's Room in Montreal*, 1846  xii
2. Emily Mary Bibbens Warren, *Nesting-Series of British Birds at Natural History Museum at South Kensington*, London, 1888  11
3. "Schema Avium Distributionis Geographicæ"  16
4. Map of Philip L. Sclater's 1858 and 1899 zoogeographic regions of the globe  22
5. "The Great Bustard in Britain"  24
6. Sketch of a hoopoe by John Gerrard Keulemans  43
7. The first issue of the *Ibis* (1859)  57
8. The Eurasian golden oriole (*Oriolus oriolus*)  65
9. A. Thornbury, "Bearded Vulture"  72
10. "Descent to Nest of Bonelli's Eagle"  74
11. "Tzelatza Valley, Morocco"  78
12. Sketch of an osprey (*Pandion haliaetus*)  83
13. Ta' Braxia Cemetery, Malta  106

MAPS

1. Route of Thomas Wright Blakiston  24
2. Routes of Andrew Leith Adams  42
3. Routes of Leonard Howard Lloyd Irby  64
4. Route of Philip Savile Grey Reid  82
5. Routes of the royal military officers  98

# Acknowledgments

The idea for this project began after I viewed an oil painting by Cornelius Kreighoff of an officer in his private quarters while stationed in Montreal in the 1840s, now housed at the Royal Ontario Museum. At the time, I was working as a collections assistant in the Department of Ornithology and had completed a thesis on the historical geographies of colonial ornithology in Upper Canada (now Ontario). The painting featured an officer, assistant surgeon Andrew Aylmer Staunton, relaxing at his desk, reading a book among the various accoutrements associated with army service in British North America, such as landscape paintings, furs, snowshoes, firearms, fishing gear, First Nations beading, books, and a bust of Shakespeare (figure 1). Animals were prominent in the room, including two live dogs, a head of a lynx, moose antlers, and at least five species of North American birds mounted on the wall (blue jay, osprey, an owl), sitting on a shelf (American wood duck), or in a glass case (scarlet tanager).

Several questions crossed my mind when I observed the details of the painting more carefully. Considering that British officers, such as Staunton, often occupied several imperial sites throughout their military careers, to what extent and in what ways did British military officers engage in ornithological activities in different parts of the British Empire? How were these activities facilitated by their postings to different sites, and did they help the advancement of their careers? How did imperial ornithologists encounter different local cultures (with different attitudes toward hunting, birds, field science, etc.) and different local natures (different sets of birds, climates, and environments)? These initial questions guided the research and my continued interest in the colonial legacies of human-environment relations in the past within the context of the nineteenth-century British Empire.

This project would not have been possible without the guidance of many individuals. First, I would like to thank David Lambert (History, University of Warwick) for his initial support of my doctoral project while in the Department of Geography at Royal Holloway, University of London. He and Alan Lester were in the process of publishing their book, *Colonial Lives across the British Empire: Imperial Careering in the Long Nineteenth Century* (2006), and he shared with me several chapters for my Social Sciences and Humanities

FIGURE 1 Cornelius Kreighoff, *An Officer's Room in Montreal*, 1846. Oil on canvas, 40.0 × 48.6 cm, Royal Ontario Museum, Toronto, Ontario, Canada.

Research Council (SSHRC) Doctoral Fellowship proposal. David would later serve as a host supervisor for my SSHRC Michael Smith Foreign Study Supplement, and as a postdoctoral mentor for my SSHRC Postdoctoral Fellowship at the University of Warwick.

As a doctoral student in the Department of Geography at Queen's University, Kingston, I benefited greatly from the co-supervision of two amazing historical geographers, Laura Cameron and Joan Schwartz. Together, they continually challenged me throughout my four years at Queen's and helped me to grow intellectually, think creatively, and embrace the process. I would not have been able to complete my dissertation without their supervision and encouragement. I also would like to acknowledge the important role that Sandra den Otter, from the Department of History, played as a doctoral committee member on my research project, and the continual support of Brian Osborne (Geography) and Bob Montgomerie (Biology), who have been my two biggest supporters.

Much of my work has been informed by the graduate courses at Queen's, such as GPHY 801 "Conceptual and Methodological Basis of Geography" and

GPHY 870 "Historical and Cultural Issues in Fieldwork" with Laura Cameron; ARTH 862 "History of Photography 1" with Joan Schwartz; and HIST 859 "Britain and the Empire, 1780–1945" with Sandra den Otter. I learned from my doctoral committee members, Audrey Kobayashi, George Lovell, and Anne Godlewska, and other faculty members, such as Warren Mabee, Beverley Mullings, Joyce Davidson, John Holmes, Jamie Linton, Betsy Donald, Mark Rosenberg, Paul Treitz, Gerry Barber, Neal Scott, Ryan Danbee, Melissa Lafreniere, and Scott Lamoureux. I would like to extend a special thank-you to the administrative staff in the Department of Geography: Joan Knox, Kathy Hoover, and Sharon Mohammed.

I am grateful for my colleagues at Nipissing University, where I have found a home both in the Departments of Geography and History and in the Masters of Environmental Studies/Environmental Sciences graduate program. My colleagues have been a source of inspiration to help me finish my book. They include April James, Jamie Murton, John Kovacs, Dan Walters, Adam Csank, Katrina Srigley, Carly Dokis, Hilary Earl, Jeff Dech, Nathan Kozuskanich, Robin Gendron, Sean O'Hagan, Jason Kovacs, and Catherine Murton-Stoehr.

A number of people have provided feedback on conference papers and journal articles that informed my book, such as David Livingstone, Alan Lester, Robin Doughty, John Tunbridge, Daniel Clayton, Felix Driver, David Gilbert, Diarmid Finnegan, Hayden Lorimer, David Matless, Bernard Lightman, Christopher Smout, Larry Sawchuk, Janet Padiak, Emilie Cameron, and Melissa Shaw. I would like to acknowledge the influence of my former professors Jeanne Kay Guelke and Suzanne Zeller, who introduced me to historical and cultural geography and the history of science while I was a master's student; and my fellow historical geography colleagues who have supported me throughout my project, including Bob Wilson, Arn Keeling, Maria Lane, Matt Farish, and Dean Bavington.

I am thankful for the expertise and assistance of Henry McGhie, John Borg, Charles Farrugia, Jim Burant, Mark Sanchez, Mark Adams, Alison Harding, Robert Prys-Jones, Paul Evans, Clemency Fisher, Nigel Monaghan, Tony Irwin, James Dean, Michael Brooke, Alastair Massey, Peter Meadows, Paul Martyn Cooper, Bob McGowan, Lorna Swift, Andrew Davis, David Reid, Zena Tooze, Andrew Reid, Sally Day, Victoria Dickenson, Kerry Patterson, Jude James, Sandy Leishman, Douglas Russell, Catriona Mulcahy, John Cortes, Damian Holmes, and Kimberley Hurni.

My project was funded by the SSHRC Joseph-Armand Bombardier Canada Graduate Scholarship, the SSHRC Michael Smith Foreign Study Supplement, the Ontario Graduate Scholarship, the Carville Earle Dissertation Research Award from the American Association of Geographers Historical Geography Specialty Group, and other internal and external grants, such as the Queen's University Tri-Council Recipient Recognition Award, and the Queen's University Dean's Travel Grant.

Finally, I would not have been able to complete my studies without the love and support of my friends and family. While at Queen's University, I developed lasting friendships with colleagues Sinead Earley, Katie Hemsworth, Heather Hall, Giselle Valazero, Kirby Calvert, Melanie-Josee Davidson, and Brendan Sweeney. I also have made new friends who helped me finish my book, including Megan Prescott, Sabrina Morrison, Laurel Muldoon, Emily Fachnie, Steven Johansson, and my family at Crossfit 705. I would not have made it without all of you. I am forever thankful to Sean Badali, Jennifer Steele, David Lutterman, Kristen Ligers, Brian and Nicole Giles, Todd Woodcroft, Sarah Borisko, Karen Borisko, Don and Mary Borisko, Heidi Bloomfield, Sascha McLeod, Eva Greer, Caroline Wetherilt, Colin Greer (brother), Thomas Greer (dad), Lillian Anderson (mom), and especially my daughter, Annsophie, to whom I dedicate this book.

# Red Coats and Wild Birds

# Introduction

For the few last decades, there has been talk of a "war" on European migrant birds in the southernmost point of the European Union (EU) and former British colony—Malta. Located in the Mediterranean Sea, Malta has long been viewed as a bridge between Europe and North Africa, with its proximity to Tunisia and Libya in the south and Sicily to the north. Each spring and autumn, thousands of European migrating birds use the Maltese Islands as a resting place for their long journeys to and from their wintering grounds in Africa. Honey buzzards, marsh harriers, hoopoes, bee-eaters, quail, and turtledoves abound, making it "a potential birdwatchers' paradise."[1]

Internationally, Malta has been portrayed as "the killing fields" of Europe and "the most savagely bird-hostile place" on the continent despite becoming a member state of the EU in 2004.[2] With the densest population in the world of bird hunters and trappers per square kilometer, migrant birds are subjected to a flurry of gunshot and traps, with many ending their lives as trophies or for pot. Malta's obligation to uphold the EU's nature conservation legislation, the "Birds Directive" (2009/147/EEC), has become a major source of conflict for the Maltese, which made it one of the top issues in the 2009 EU elections when I was there in the spring of that year.[3] Maltese bird hunters and trappers espouse their traditional and legal rights to engage in spring hunting, using the UNESCO Convention on the Protection and the Promotion of the Diversity of Cultural Expressions to support their claims.[4]

As part of Europe's shared "natural heritage," the illegal bird hunt has enraged northern European environmentalists, who condemn publicly the killing of protected migratory birds. Framing the hunt as a mass slaughter, hundreds of bird volunteers, many from the United Kingdom, descend on the islands every spring and fall to stop the butchery. According to popular British nature writer Michael McCarthy, writing in 2008: "It is Europe's worst and most senseless wildlife massacre, resulting in the annual death of thousands of birds about to breed all over the continent. But this spring, conservationists are heading for the Mediterranean to help end it."[5] Such impassioned declarations have ignited physical confrontations among bird protectionists and bird hunters in the Maltese countryside, illustrating the intensity of the war on European migrant birds.

Moreover, Malta's so-called unnatural relationship with birds has been put into sharp relief in comparison to Britain's other previous Mediterranean colony—Gibraltar. Once a monument to empire, the British overseas territory is now promoted as a model of nature conservation and ornithological study in the Mediterranean. Here, migrant birds are protected and studied, in contrast to Malta or Cyprus, where, according to McCarthy, "only one fate would await these birds: the pot. Here on the 'island' of Gibraltar they are sacrificed twenty minutes of freedom in the cause of ornithology and were sent on their way."[6]

While some people have claimed that the EU is another form of imperialism now imposed on the Maltese, what is missing from this understanding, I believe, are the ways in which bird protection in Malta, the production of the Maltese "pothunter," and environmental ideas of British migrant birds and semitropicality are rooted in part in Britain's imperial past in the Mediterranean region.[7] By the mid-nineteenth century, Malta, in addition to Gibraltar, served as a transimperial station that connected Britain to Asia, as well as British North America, West Indies, and Africa, through the flow of military manpower, commodities, information, and bodily experiences across the empire. Known as the "empire route" and the "artery of empire", the "material chain" of military stations in the Mediterranean provided "the shortest route to India" and formed "the spine of prosperity and security" of the British Empire, especially with the opening of the Suez Canal in 1869.[8] The strategic position of Gibraltar and Malta also facilitated increased access of Britain's army to different parts of the empire, as Britain's involvement in the "small wars" in India and Africa and its imperial interests in China increased the demand for troops.[9] British imperial policy with regard to the Mediterranean therefore reflected the role of this strategic route to India from Britain.[10]

In order to trace some of these trajectories, this book focuses on the nineteenth-century British Mediterranean, when the region emerged as a crucial location for the security of British trade routes to India and South Asia, and when the acquisition and maintenance of Britain's global empire depended on the efficiency and presence of military manpower stationed at key sites in the Mediterranean region such as Gibraltar and Malta. As Philip Howell has stated, the British Mediterranean has often been left out of the histories of the British Empire and "deserves greater notice,"[11] despite significant work on the region by environmental historians and historical geographers.[12]

Britain's imperial presence in the Mediterranean region was contingent on other competing empires (i.e., French, Spanish, Ottoman, Italian, Russian, German) in its attempts at maintaining control in the region and bringing North Africa into its "informal empire," described here as the links fostered by trade, investments, diplomacy, or scientific networks that drew new regions into the world-capitalist system.[13] Britain relied on Islamic regimes for basic sustenance at its garrisons, especially at Gibraltar, where Morocco was the main supplier of bullocks during a time when British military reforms included a change in diet from salt meat to fresh meat.[14] In Tunis, British officials depended heavily on bullocks, sheep, fruit, and vegetables in return for Manchester cottons, Sheffield knives, London pickles, sauces, and tinned meat (Manai 2006, 367–68). British imperial policy in the Mediterranean thus engendered its own unique geopolitics and practices between Protestant Britons, European Catholics, Mediterranean Jews, and North African Muslims, reflecting in part the limits to British imperial power in the region (Colley 1992, 132).

Furthermore, Britain experienced significant changes at home and abroad with shifting geopolitical circumstances (i.e., the Crimean War, the "Indian Mutiny," the "Jamaican Rebellion," the Franco-Prussian War, the "Scramble for Africa"), which altered natural and national boundary-lines and spatial orderings of empire. Some believed Britain's decline in military authority was a result of a regimental system that promoted "a culture of gentlemanly amateurism,"[15] sparking changes in the British army with the Cardwell-Childers Reforms (1870–81) and the Royal Commission on the Defence of British Possessions and Commerce Abroad (1879–82).

At home, Britain witnessed rapid industrialization, urbanization, mechanization, and changes in the rural landscape, creating class tensions and changing the ways in which people interacted with urban and rural landscapes. New technologies of travel and communication (i.e., the railway, the telegraph, the steamship) helped to shift perceptions of time and space, creating a "crisis in identity" that resulted in a devotion to more localized memories of place, nation, and region.[16] The perceived degenerative effects of progress from industry, capitalism, and social mobility also resulted in the dismantling of "fixed hierarchies, places, and temporal trajectories," creating anxieties over the loss of "our [British] world."[17] These transformations resulted in the rise of national heritage preservation movements in Britain, which advocated for preservation based on "cultural value," and the idea of the countryside as a repository of a way of life that required protection, which, this books argues, extended to the migrant birds associated with these landscapes.[18]

The British Mediterranean was where British army officers started to think about "modern" ideas of bird migration from Europe to Africa, as well as the boundary between the temperate (Europe) and tropical (Africa) worlds. It was in Gibraltar that English naturalist Mark Catesby's (1682–1749) military brother collected birds when stationed there in the 1740s and sent specimens to British naturalist George Edwards, who wrote about bird migration.[19] Considering the scale of Britain's global empire, the British armed forces composed the largest group of bird collectors across the globe in their zeal for "sporting zoology."[20] Their privileged positions in colonies abroad allowed them access to places that were off limits to the regular traveler and naturalist, helping them build up vast collections of bird skins, eggs, and nests, while their transient movements across different imperial sites created opportunities for comparison of birds, peoples, and landscapes. Considering how the collecting practices of British military officers were integral to the establishment of many natural history collections in Britain, their natural history collections, as well as the birds commemorated with their names, present historical and cultural meanings intricately linked to identity, colonial expansion, and empire. Thus, the flow of wild birds provided a tangible link from Britain to its colonies in the Mediterranean, Africa, and Asia.

Surprisingly, studies on the history of ornithology have also overlooked the contributions of British military officers and their role as men at the forefront of empire. E. G. Allen's seminal work on the history of ornithology includes some biographies of British military officers who contributed to the field, such as Blagden and Sabine of the Royal Artillery.[21] Similarly, Barbara Mearns and Richard Mearns highlight the lives of similar individuals who collected birds in different regions of the British Empire.[22] Yet, their works remain popular histories of ornithology, which require further analysis into the ways imperial and gendered positionalities shaped military ornithological practices and ideas across the British Empire.[23]

Research on ornithology as a military practice has emphasized the ways ornithological fieldwork and military cultures have facilitated territorial conquest and maintenance in empire- and nation-building projects. Edgar Hume's *Ornithologists of the United States Medical Army Corps* (1942) discusses the numerous military officers of the United States Medical Corps who surveyed, collected, and observed wild birds during American expansion and settlement in the western United States in the nineteenth century. As a form of nation-building, these officers pursued ornithology largely in association with the Smithsonian Institution, which housed the collections of stuffed birds and ornithological data collected by members of the Medical Corps.[24]

Roy MacLeod demonstrates how ornithology sustained American imperial and territorial interests as a form of empire-building in the Pacific Ocean during the Cold War. Knowledge of Pacific birds became a national and military interest in order to produce a comprehensive environmental understanding of the area through the Smithsonian's Pacific Ocean Biological Survey Program.[25] As Helen MacDonald has shown, these practices extended to civilian amateur scientists who observed birds to help monitor Britain's coast during World War II. Amateurs were trained to look not only for birds but also for enemy planes approaching the British coastline. According to MacDonald: "Birds were more than just birds: they were loaded with symbolic effect increasingly derived from versions of national identity." These works provide important links to the ways military and ornithological practices have helped shape and maintain territorial boundaries.[26]

Within the discipline of geography, the rich and exciting work on the geographies of science has neglected the role of birds in the accumulation of geographic knowledge and empire-building.[27] Viewed as a noninstrumental science with little economic value, field ornithology has remained in the background of critical studies of empire and science.[28] However, as historical geographers have highlighted, birds have connected landscapes across space and time, shaping people's identities and experiences of place.[29] One example is Robert Wilson's work on the practices of migratory bird protection at the continental, regional, and local scales. By focusing on the U.S. Fish and Wildlife Service's wildlife refuges, Wilson follows the Pacific Flyway (a migratory route) through disputed and changing human geographies of North America, which included disgruntled farmers and enthusiastic sportsmen. Wilson pays particular attention to place and scale in histories of environmental change, recognizing space as dynamic, as well as the relationship between scales rather than focusing solely on the local, regional, or national. He also illustrates how the Pacific Flyway connected a series of landscapes from the Arctic to the Canadian prairies, California, and western Mexico.[30] Because nothing comparable exists for the Mediterranean region, it is through these works that I situate my book, which traces the politics of migrant birds in the Mediterranean to when the region was an integral part of Britain's global empire through imperial defense.

The book works by juxtaposing the biographies of two particular groups of transient and mobile figures (British military officers and migratory birds) in a particular region (the British Mediterranean) as a means of capturing the circuitry of empire in the production of British imperial knowledge, territory, and identity. Here, I follow and extend the work by historical geographers David Lambert and Alan Lester on biography and geography to include a network approach to uncovering the life paths or "life geographies" of individuals who contributed to the British Empire in the long nineteenth century.[31] As Lambert and Lester note, "Military figures are one category that would benefit from future critical research."[32] This book therefore focuses on four individual officer-ornithologists who intersected in the British Mediterranean in the nineteenth century: Thomas Wright Blakiston (Royal Artillery), Andrew Leith Adams (64th and 22nd Regiments), Leonard Howard Lloyd Irby (90th and 74th Regiments), and Philip Savile Grey Reid (Royal Engineers), all of whom served and collected birds associated with the Mediterranean region, such as the Crimean "theater of war," the "Rock" of Gibraltar, the Aldershot garrison, the Valletta market in Malta, the Spanish countryside, and "the wilds" of North Africa.

Of critical interest are the ways in which imperial careering shaped transimperial military identities, but also their roles as agents in territorial control with guns and military force. These men circulated across the circuitry of empire that connected different places and diverse peoples in (and beyond) the British Empire. With this approach, they demonstrate the "movement of networks of knowledge, power, commodities, emotion and culture that connected the multiple sites of the empire to each other, to the imperial metropole and to extra-imperial spaces beyond."[33] Each site—colony, port, battlefield, troopship, "the field," home—within the imperial network fostered "'its own possibilities and conditions of knowledge.'"

As British military officers came into contact with distinct local indigenous cultures and natures, they also occupied successively different postings, involving them in translocal activities, networks, and military cultures.[34] It is through these embodiments in different imperial places that British military officers (re)shaped ideas about masculinity, whiteness, class, and Britishness, which impacted the way credibility and authority were negotiated in the field. This book follows the work on embodiment and performativity to examine the material interactions at the level of the body practice, and to link identity formation through particular practices with landscapes through notions of gender, class, and whiteness.[35] Place is therefore viewed as residing in people "both embodied and narrated and is, as a consequence, often highly mobile: places travel with people through whom they are constituted."[36]

While most works on the British Empire have emphasized the importance of human actants in shaping territoriality, this book recognizes the materialities of "place" and the affective relationships between people, things, and the "more-than-human" world.[37] The Mediterranean region, in particular, was considered a major site for the migration of birds to and from Europe and Africa. The bodies of migratory birds therefore hold particular significance in this study as they intersected with the lives of transient British military officers.[38] Both human and animal bodies are "in perpetual flux. . . . These beings do not exist at locations, they occur along paths."[39] This notion of "place" suggests a meeting point of "mobile lives" and narratives as encounters, forming convergences and configurations of trajectories, all with their own temporalities.[40] But these encounters also imply the "non-meetings-up, the disconnections and the relations not established, the exclusions," which all contribute to the formulation of "place."[41]

A unique aspect of this book therefore is the inclusion of the lives of birds in these networks and in their wider biogeographies connecting landscapes and places beyond borders and boundaries.[42] This includes integrating the life geographies and migratory routes of particular bird species—such as the great bustard (*Otis tarda*) in the Crimea, the golden oriole (*Oriolus oriolus*) at Gibraltar, the hoopoe (*Upupa epops*) in Malta, and the osprey (*Pandion haliaetus*) at Aldershot—in short avian vignettes at the beginning of each chapter. These avian biographies will highlight the ways in which migratory birds intersected with imperial lives in particular places connected to the Mediterranean.

Chapter 1 situates these officers by providing context for the contributions of military officers to the development of field ornithology from the traces and material remnants of their bird collections and specimens housed in museums across the British Empire, especially in Britain. Untangling the avian imperial archive explores how transimperial careers can be written using not only textual sources (e.g., biographies and personal correspondence) but also traces and artifacts of material culture, specifically bird skins as part of the avian imperial archive. By unraveling the avian imperial archive, the contributions of British military officers to the emergence of the field of zoogeography—a branch of biogeography concerned with the distribution of animal species across the globe—are put into sharp relief, illustrating the multiple avian-human entanglements in different parts of the British Empire, including in the Mediterranean. As both a fantasy of empire and a reflection of transient lives, avian scientific specimens in historical geographic research enrich our understanding of the intersections between science, empire, and the military, as well as the role of the researcher in these networks.

Chapter 2 examines the production of the scientific war hero in British military culture in the mid-nineteenth century, with an emphasis on the Crimean War (1853–56) as an important event in securing Britain's ascendency over Russian aspirations in the Mediterranean region, and in the emergence of the military-scientific hero. The chapter also highlights the military-scientific hero as a product of conducting fieldwork in the Crimean theater of war and collecting specimens as scientific trophies of war for a British audience at home. Here, I focus on Ordnance officer Captain Thomas Wright Blakiston, Royal Artillery, who collected numerous birds while serving with his regiments, published works in the *Zoologist*, and sent specimens to British museums, including the Museum of the Royal Artillery Institution at Woolwich.

Set in Malta, chapter 3 follows the military medical career of Andrew Leith Adams, military surgeon with the Twenty-Second Regiment of Foot, whose military and scientific networks and travels to northern India, Malta, Egypt, and New Brunswick, British North America, helped him to conceive ideas of tropicality, semitropicality, and the temperate. To Adams, temperate martial masculinity was both a physical and mental state and a climatic zone important in the maintenance of a British military career across the British Empire. His ornithological investigations also allowed him to contemplate the zoological connectivity between Europe and North Africa.

Chapter 4 analyzes the ways in which ideas, practices, and performances of ornithology helped to sustain territorial maintenance and British imperial place-making in the Strait of Gibraltar by focusing on the work of Lieutenant Colonel Leonard Howard Lloyd Irby (Ninetieth and Seventy-Fourth Regiments). Located in the Mediterranean, the island-like territory of Gibraltar emerged as a strategic geopolitical position in the preservation of the British Empire and served as part of the "artery of empire" that linked Britain to India. It was also an important landmark in the British imagination as a result of the Great Siege (1783) and its resonance for Horatio Nelson in the Napoleonic Wars. This chapter demonstrates how narratives of wild birds and scientific performances surrounding the British military officer attempted to legitimize Gibraltar as an imperial, noble, and masculine pillar of empire, and to extend imperial interests into Morocco and Tangier.

Chapter 5 investigates how, back "home" in Britain, British military officers' production of ornithological knowledge in the British Mediterranean helped reformulate notions of nation and "British birds," especially as officers often returned to Britain after tours of duty or retirement. Informed by works on "homeland," "home," and landscape and identities, this chapter centers on the career of Captain Philip Savile Grey Reid, Royal Engineers, to examine the various domestic spaces, including the military base at Aldershot, where British military officers engaged with ideas and practices of ornithology. Many officers who served in Gibraltar and Malta contributed to the development of British ornithology by publishing books or assisting with the arrangement of British birds at museums. For example, Irby prepared a *British Birds: Key List* (1888) and assisted in the formation of the life groups of British birds in the British Museum (Natural History), while Reid obtained birds for the nesting groups of British birds series in the galleries at the British Museum. Their published works, exhibitions, and lectures helped to shape ideas about domestic birds for an audience interested in birds and bird protection at the end of the nineteenth century.

The concluding chapter, chapter 6, demonstrates how the accumulation of their avian collections and documentation served as an ideological force in imagining control over universal knowledge and, in turn, the British Empire and its territories, as officers studied birds as part of surveying, mapping, and surveillance. It analyzes how military ornithologists encountered different local cultures (with different attitudes toward hunting, birds, and field science) and different local natures (with different climates, avian populations, and environments), and how imperial knowledge was contingent on local networks and of different trajectories across the British Empire.

As this book reveals, the British Mediterranean was not a bounded region, a "single geographic unit,"[43] or a fixed "culture area," but was entangled in global "circuits" of empire and biophysical processes.[44] As Lucien Febvre once noted, the Mediterranean has been created by the "movements of men, the relationships they imply, and the routes they follow."[45] To Febvre's definition, we can add birds, winds, trees, and other nonhuman actors that constituted these assemblages in different times and places.

CHAPTER ONE

# Red Coats and Wild Birds across the British Empire

In 1882, the Natural History Museum at South Kensington opened its doors to a new exhibition, *Nesting-Series of British Birds*. Located in the first room of the modern building, the display showcased a variety of avian species found within the British Isles (figure 2).[1] The exhibition was based on the collection of groups of birds, nests, and eggs in their natural state from England, Scotland, and Ireland, with each species (159 in all) displayed in separate glass cases and habitats in the museum's bird galleries. In most instances, the birds and their nests were "exhibited with the actual tree, rock, turf and other support which was found with them," as well as the details of the collector, the county, and the date.[2] Such species included the "Robin Redbreast," one of Britain's "most familiar and characteristic resident species . . . where legendary associations and its fearless nature have combined to make it a general favourite."[3]

A number of British military officers contributed to the nesting series (figure 2), including a little tern by Willoughby Verner and a puffin by Colonel Paget Walter L'Estrange, Royal Artillery. L'Estrange was known for collecting birds during his military service abroad and gifting his bird skins to the Royal Artillery Institution at Woolwich.[4] Lieutenant Colonel Leonard Howard Lloyd Irby prepared a *British Birds: Key List* (1888) and assisted in the formation of the life groups of British birds in the British Museum (Natural History), while Philip Savile Grey Reid obtained birds for the nesting groups of British birds series in the galleries at the British Museum. Others included Captain George Ernest Shelley, Grenadier Guards, who became a leading expert on African ornithology after serving on the Gold Coast and South Africa, and exploring in Central Africa.[5]

Army officers such as Irby and Reid were the most numerous members of the British Ornithological Union and assisted in the accumulation of knowledge of birds across the British Empire.[6] These military men also sent bird skins to the Natural History Museum from these regions, which made up the ornithological collections in the museum's back rooms. In the 1880s, Richard Bowdler Sharpe (1847–1909) of the British Museum urged the War Office to encourage and reward those officers who took advantage of "their opportunities for increasing scientific knowledge."[7] Today, many of these

FIGURE 2  Emily Mary Bibbens Warren, *Nesting-Series of British Birds at Natural History Museum at South Kensington*, London, 1888. Natural History Museum.

specimens continue to form part of the collections at the Natural History Museum, as well as other natural history museums across the United Kingdom and North America.[8]

Dead birds circulated across the British Empire in a myriad of ways. Stuffed birds were sent as gifts in the slave trade; as part of the discourse of discovery and exploration; as songsters in aviaries in gardens; as treasures in private museums; as colonial objects at world fairs; and as scientific specimens in the emergence of the field of ornithology, the scientific study of birds. The collections and documentation of avian specimens bore witness to the "circuitry of empire"[9] that extended across formal and informal parts of empire, including missionary work in the South Pacific, coffee plantations in Latin America, fur trade in British North America, timber commerce in Atlantic Canada, and military service in the British Mediterranean. Here, the avian imperial archive reflected the "movement of networks of knowledge, power, commodities, emotion and culture that connected the multiple sites of the empire to each other, to the imperial metropole and to extra-imperial spaces beyond."[10] It also brought into sharp focus the transimperial lives of the collectors and the relationship between empire and home.

Within the context of the British army, imperial expansion provided opportunities for many British military officers to pursue natural history in colonies abroad; in fact, the British armed forces accounted for the largest number of bird collectors in the British Empire. Commissioned officers emerged as ideal observers and collectors in Britain's formal and informal empire as they documented, listed, classified, and narrated their presence in the field. As evident in many British museums, the collecting practices of British military officers were integral to the establishment of many collections in Britain and contributed significantly to ornithological knowledge of different regions of the globe. Considering how the contributions of military officers to the development of field ornithology and zoogeographic knowledge of birds can be found in the documented traces and material remnants of their bird collections housed in Britain and its former colonies, how does one use them in tracing the lives of avian specimens and British military men?

This chapter examines the ways in which transimperial careers can be written not only with textual sources (e.g., journals, personal correspondence, published works) but also with traces and artifacts of material culture, specifically avian specimens as part of the "avian imperial archive." In so doing, it brings a material-cultural methodological approach to bear on the life geographies of British military officers, who accumulated a variety of material culture, artifacts, experiences, and ideas from their transimperial travels, all of which made up the "avian imperial archive."[11] In employing this term, inspired by Thomas Richards, this chapter inserts military ornithological sketchbooks, diaries, albums, and avian specimens into the "imperial archive."[12] Like the imperial archive of which it is part, the "avian imperial archive" acted as "an ideological construction for projecting the epistemological extension of Britain into and beyond its empire."[13] As Emyr Evans has demonstrated, the material culture of the Royal Artillery and Royal Engineers in Ireland consisted of material reports, maps, diaries, and books, which Evans termed "the temple of facts bequeathed by the bureaucracy of empire."[14] To his list we need to add natural history specimens, as the accumulation of avian geographic knowledge by British military officers also reflected the "service of state and Empire."[15]

The "bodies of animals" have been, and continue to be, sites of political struggle over the construction of cultural difference and maintenance of dominant ideologies. They have been used to racialize, dehumanize, and maintain power in several ways.[16] As tangible "things" from the past, the bodies of dead birds can be used as a site for an analysis of the intersection between British military culture and ideas and practices of ornithology in the nineteenth century, reflecting the ideological and collecting practices of a network of "servants of empire" dedicated to the scientific study of avian lives. According to Donna Haraway, "Behind every mounted animal . . . lies a profusion of objects and social interactions among people and other animals, which in the end can be recomposed to tell a biography."[17] Works in the history of science have therefore shed light on the cultural biographies of museums, natural history collections, their collectors.[18] With a focus on the specimen as a scientific object, these studies reveal complex networks, biographies, and institutional histories as objects circulated through colonial, imperial, and private and public spheres.[19]

Geographers have devised creative ways in which to interpret the various "fragments" of the archives that often lie beyond the traditional sources of the historical record and, in this context, the scientific record.[20] These interventions help to conceive the stuffed bird specimens in my study not as "discrete entities" but as material forms "bound into continual cycles of articulation and disarticulation" that have the potential to reveal other histories, which include the "lived culture" of the animals in question.[21] The ornithological specimens also reveal the historical geographies of scientific practice, military culture, and mobilities, linking animals to critical military geographies.[22]

A focus on the "lives of specimens"[23] seeks to show how bird skins reveal networks of different actants (human and nonhuman) in the formulation of ornithological knowledge and zoological practices, including the transimperial movements of British military officers in different environments and cultures, and their connections to colonial officials, patrons, local assistants, and those who opposed them. By tracing the lives of military men through their bird collections, this approach takes into account the "life geographies" of soldier-naturalists. It also draws attention to the need for examining the different sites that scientists occupy during their careers and how these "places" influence scientific knowledge-making.[24] This research focuses primarily on the process of making science and such aspects as social context (structure) and personal creativity (agency); it concentrates on "lives lived" and not merely on the final results of the scientist.[25]

As an approach, life geography need not ignore the birds themselves, their own movements and relations with other creatures—and their mysteries.

Instead, this approach may attend to interactions with avian species as a means to include the lives of birds (alive and dead) as actants in transimperial networks and as living beings in their wider biogeographies connecting sites beyond borders and boundaries. As Adrienne Rich has articulated: "The Great Blue Heron is not a symbol . . . it is a bird, *Ardea herodias*, whose form, dimensions, and habits have been described by ornithologists, yet whose intangible ways of being and knowing remain beyond my—or anyone's—reach."[26]

In addition to reading the avian specimens solely as texts or representations, this approach focuses on the material practices that helped produce their meaning and epistemic authority as scientific facts in the production and circulation of ornithological knowledge.[27] This requires familiarity with the various field ornithology practices—collecting, stuffing, labeling, travel writing—in order to understand the ways in which a network of British military officers and naturalists helped to translate avian specimens into scientific facts.[28] British military officers not only came into contact with distinct local cultures and natures but also occupied successively different postings, involving them in translocal activities, networks, and military cultures. Here were the "embodied interactions that create[d] embodied subjectivities *and* standardized facts" in the production of colonial knowledge.[29] As the avian imperial archive illustrates, the "traffic in objects" involved patrons, collectors, local informants, taxidermists, poulterers, agents, and museum officials, revealing networks of exchange, coercion, and the hybridity of knowledge creation.[30] A special attention to biography and practice therefore helps to discern the uses of avian specimens in fieldwork, "with all the complications and translations entailed."[31]

With increasing emphasis on technologies of inscription (writing, documenting, mapping, photographing, taxidermy), we can recast ornithological knowledge and objects that circulated as "immutable mobiles" through a "circuit" or "web" of networks that shaped and reshaped places across the nineteenth-century British Empire.[32] The mobility of geographic knowledge currency required "centres of calculation" to standardize collecting practices, coordinate flows of information, and manage networks of communication in order for information to be useful for further explorations.[33] Within the British imperial context, such centers included the British Museum in London, which served as a "nerve center of all possible knowledge" under the supervision of the state.[34]

While viewed as a "non-instrumental" science with little benefit "on the pretext of economics," British military ornithology was very much part of imperial region-making, surveillance, campaigning, and military strategy.[35] As Janet Browne has discussed, theories of zoogeographic distributions were "equally grounded in geopolitical concerns," and these "too reflected the ethos of empire.[36] One of this book's main contributions, therefore, is to show how British military officers contributed to the field of zoogeography—a branch of biogeography concerned with the distribution of animal species across the globe—which involved the mapping of the six zoogeographic regions of the world and their subdivisions. Here, zoogeography was coconstituted of British imperial geopolitics.[37]

The accumulation of bird skins, eggs, nests, field journals, and sketchbooks by British military officers stationed in the Mediterranean helped to provide evidence for the process of abstraction and the production of zoogeography—a branch of biogeography concerned with the distribution of animal species across the globe.[38] By the mid-nineteenth century, the study of the geographic distributions of passerine birds allowed for the conceptualization of the six zoogeographic regions of the globe by British naturalists, which contributed to the mapping of "the most natural division of the earth's surface into primary kingdoms or provinces," including their "secondary divisions" or subregions.[39]

In 1858, Dr. Philip L. Sclater figured verbally and diagrammatically the six zoogeographic regions using the distribution of birds, which included the boundary between European and African zoologies in the Mediterranean region (figure 3). According to Sclater, the Palearctic region encompassed Europe, northern Asia, and "Africa, north of the Atlas, along the southern shores of the Mediterranean," which appeared "to belong to Europe zoologically, and not to the continent to which it is physically joined."[40] For Sclater, North Africa appeared to be a place where some bird species seemed to be "slightly modified representatives" of the European ones.[41] British naturalist Alfred Russel Wallace extended Sclater's region to "all Africa north of the great Desert, for I think none of the peculiar forms of Tropical Africa are found there."[42] Wallace described the Sahara Desert as a boundary between the Palearctic and Ethiopian regions, writing that it "should be given to neither," as it is "certainly quite as unproductive of animal life as the sea, perhaps more so."[43] Although Sclater's zoogeographic boundaries were based on the idea of multiple centers of creation, his emphasis on regions helped Charles Darwin and Wallace to conceptualize evolutionary theories of a single origin.[44]

In order to map empirically the limits of the world's zoological provinces, Sclater appealed to field collectors such as British military officers to engage

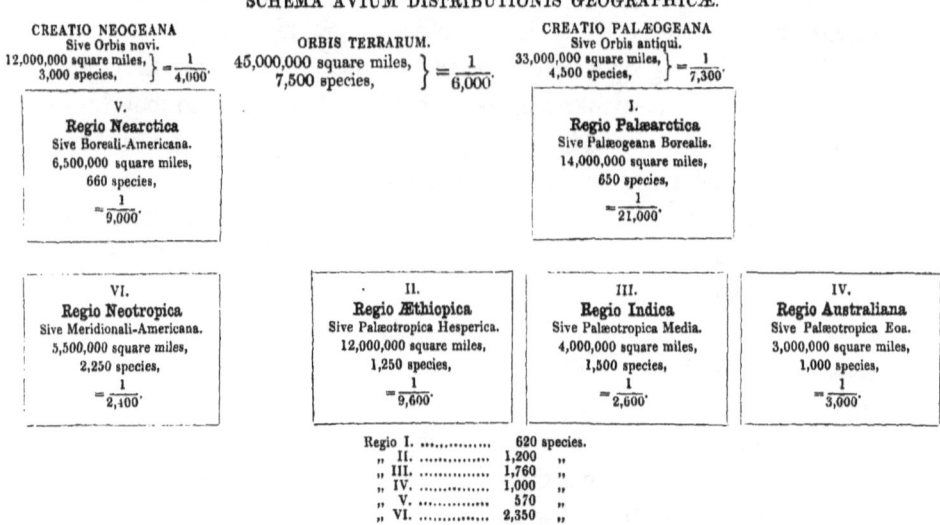

FIGURE 3 "Schema Avium Distributionis Geographicæ," in Sclater, "On the General Geographical Distribution of the Members of the Class Aves," 145.

in the accumulation of accurate "information concerning the families, genera, and species of created beings—their exact localities, and the geographic areas over which they extend" in order to map accurately the "ontological divisions of the globe," including in the Mediterranean subregion.[45] Sclater, through his involvement with many leading scientific societies such as the Zoological Society of London and the British Ornithological Union (BOU) and as founder and editor of the BOU's journal the *Ibis* (est. 1859), emerged as a "centre of calculation" through a network of colonial officials, travelers, and British military officers, who contributed to the mapping of "the most natural division of the earth's surface into primary kingdoms or provinces."[46] The *Ibis*, in particular, concentrated primarily on ornithological writings from the "Mediterranean lands" in its first few decades.[47]

British naturalists knew the value of enlisting the services of army officers stationed in the Mediterranean for information on bird life and avian migrations. As early as the 1740s, naturalist George Edwards encouraged military Englishmen in Gibraltar "who reside there" to document bird migrations, a tradition that continued well into the mid-nineteenth century, when trained officers such as E. F. Becher, Royal Artillery, traced "the connection of the rock [Gibraltar] with Africa" through the study of birds.[48] As the link between the British army and scientific patronage became an "elaboration of identity,"[49] close associations with leading scientific societies such as the Royal Geographical Society, the Royal Society, and the British Association for the Advancement of Science (BAAS) fostered an emphasis on the accumulation of geographic knowledge and gathering accurate information for military intelligence, campaigning, and territorial expansion, illustrating how military connections often overlapped with and fostered scientific networks.[50] British military geographers thus cultivated a military avian landscape perspective through "nature's intelligibility" in the form of maps, sketches, photographs, and natural history specimens influenced by their imperial martial subjectivities.[51]

The military surveillance of colonial and avian landscapes through bird collections, travel writing, and sketches in the Mediterranean created a network of "panoptic power"[52] that facilitated the mapping of the zoogeographies of the Mediterranean, extending the European or Palearctic zoogeographic boundary-line into North Africa, often overlapping with British geopolitical interests in the region (see chapters 3 and 4).[53] In the nineteenth century, Britain strove to figure the Mediterranean Sea as an "English Lake,"[54] a domestic region to ensure its role as a great power in continental Europe, to maintain supremacy over the trade route to India and East Asia, and to retain political influence in North Africa.

British military officers stationed in Gibraltar and Malta thus helped to bridge the continents of Europe and Africa through the collection, comparison, and classification of avian observations, lists, specimens, and publications in formal and informal empire. Once an "immutable mobile," imperial bird skins circulated to private and public museums for authentication and entered into "cycles of accumulation" for demarcating precisely the boundary-lines between regions "contradicting, affirming and refining the boundaries of Sclater's regions and the subregions within them."[55] Lieutenant Colonel Irby's collection of birds from Gibraltar and southern Spain, for instance, was presented to the British Museum in several separate donations throughout the 1870s and 1880s.[56] Irby also corresponded personally with Sclater, sending him specimens from the Mediterranean.[57]

In order for ornithological knowledge to become useful, British military officers established techniques of trust as the first step in gathering the world.[58] British naturalists emphasized the accumulation of species lists in specific geographic locales by appealing to "a sizable network of observers."[59] These lists followed the naming conventions laid out by the BAAS in 1842,[60] which ordered the avian class with Linnaean nomenclature and created a "grid" that took into account which bird species belonged to which zoogeographic region; in this way avifaunal regions were presented as "bounded, determinate, and therefore—in principle—countable."[61] The Linnaean classification system provided a medium to convert natural history into the "visible," as language and natural phenomena became one, and the universal standardized vernacular knowledge.[62] As Matt Hannah has argued, these "grids of specification" produced "coordinate systems on which to locate the objects of a discursive formation and their features," reflecting the standardization of knowledge and the establishment of categories in the nineteenth century.[63]

The "field" allowed British military officers to make important contributions through their "on-the-spot" observations and collections in different colonial locales.[64] From the time of Linnaeus, traveling naturalists could gain the "increasing prestige and authority of internationalized expertise,"[65] spurring debates between closet/armchair and field naturalists over who held the most authoritative perspective on science and nature.[66] Field naturalists believed in viewing specimens in the "wild" and in their natural habitats, while museum naturalists drew from knowledge "not from passage but from immobility,"[67] thus relying on a vast collection of specimens for comparison. Traveling naturalists, in contrast, endeavored to emulate the fieldwork of Alexander von Humboldt, who attempted to move beyond "the classification systems of the 'miserable archivists of Nature'"[68] by traveling to distant lands and using his body as an instrument of science in documenting empirical knowledge.

The practice of taxidermy helped materialize "the natural order" of things, where "natural form tended most easily to assume the form of logic" through the arrangement of single specimens or "type specimens."[69] As representatives of "the organisms (and taxa) from which they derived," the boundaries of species could change with new discoveries and classification schemes, but their names could not.[70] Type specimens safeguarded "the permanence of the name,"[71] and in turn the imperial legacies of British military officers named in their honor.[72] As early as the 1770s, Thomas Davies (1737–1812), Royal Artillery, described for the president of the Royal Society the "modes for preserving such things," with instructions on the type of "shot" and how to stuff a bird properly.[73] He appealed to the "officers in the Majesty's Army and Navy," who had the advantage "by means of their profession of visiting different parts of the world," to form a collection of birds.[74] Within the context of ornithology, an impressive bird collection with rarities was a recognized form of capital in the appropriate circuit, with significant exchange value and an indispensable prestige function that could advance its owner into the ranks of the professional sciences.[75]

Alfred Russel Wallace urged the "modern naturalist" to amass a "geographical collection" of specimens and to focus on the relations between animals and the environment rather than on anatomy or classification.[76] Influenced by von Humboldt, naturalists translated spatial groupings of natural phenomena into numerical terms similar to population surveys involved in "enumerating political states."[77] The numbers of species and genera were counted, and simple correlations and proportions were calculated in specific regions in order to assess the number of individuals in different groups. Great emphasis was placed on the "locality" of the avian observation and collection as a key component in justifying knowledge claims. The emergence of the "type locality"—the specific geographic location where the collector obtained the type specimen—inevitably took on significance in the mapping of zoogeographic regions and the preparation of specimens collected in the field.[78]

In his "Hints for Preparing and Transmitting Ornithological Specimens from Foreign Countries," Sir William Jardine (1800–1874) emphasized how collectors should include a label with "a number, the locality, the date, and the sex" of each bird collected, with a number that "should refer to a memorandum book, where, when possible, all extended remarks that can be collected should be inserted," which included vernacular names.[79] Jardine stated that "a collection, accompanied by such memoranda, would be worth, in the market, double, if not triple to that of one indiscriminately made, even though the specimens were in finer preservation."[80] The collection of birds thus became "allied to travel narratives" as traveling and collecting became systematizing activities,[81] especially as many British military officers were members of the Royal Geographical Society, an organization that published works on "hints to travellers" for the accumulation of accurate geographic information.[82]

The recording of "on-the-spot" observations was made possible through portable field journals, sketchbooks, scrapbooks, and albums, now often separated from the specimens and housed in the archives. Journals were essential in recounting fieldwork experiences especially when publishing travel narratives. Andrew Leith Adams used his India journal to preserve "the objects in the order they appeared to me, and attempted to describe the scenes and circumstances with which I was brought in contact as minutely as the incidents of travel would allow, and in a belief that my jottings by the way would add zest to the drier descriptions of animals."[83] In some cases, notebooks and collections were lost in transit as British military officers moved frequently to different colonial stations. Lieutenant Colonel Irby lost all of his Crimean notes when he was shipwrecked with the Ninetieth Regiment (Perthshire Volunteers) on HMS *Transit* in the Strait of Banca, "so all the information is from memory only."[84]

Sketching and photography also helped to justify locality in "the craft of empirical representation" in the field.[85] Officers, especially those with the Royal Artillery and Royal Engineers, were trained in watercolors and topographical sketching.[86] Photography was a new mode of encountering the world in the nineteenth century, and its power was harnessed in a number of ways: to establish imperial control, to extend imperial authority, and to show presence at a particular site. Collections of British military albums therefore represented sites where "facts" could be visually stored, classified, reinterpreted, and disseminated, creating an imperial geographic imagination of colonial stations.[87]

Furthermore, the collection of avian geographic knowledge extended to the accumulation of birds' eggs and nests, which reflected changes in what counted for evidence of nesting habits. Oology, the scientific study of eggs, and nidification, the scientific study of nesting, emerged as subdisciplines in ornithology. As Captain Thomas Wright Blakiston of the Royal Artillery counseled, "The careful identification of an egg is too of the utmost importance," and bird specimens without their eggs "are of no value whatever scientifically."[88] Officers were encouraged to shoot the "female" bird "from the nest—and to preserve her skin—making the eggs to correspond with it."[89] Once the eggs were collected, they were to be blown out "at the side from a single hook," and notes should be written on them.[90] Blakiston concluded: "The scientific value of eggs thus carefully collected and written upon is not to be overrated. Their identity is thus secured for all future time."[91]

However, the production of British military ornithological knowledge and collections in the Mediterranean region emerged as "a negotiation of local knowledges of conjunctural context" in both southern Europe and North Africa (see chapters 3 and 4).[92] As field ornithology relied strongly on the contributions of local assistants, translators, and agents in the accumulation and knowledge of avian specimens, the participation of Gibraltarian, Spanish, Maltese, and North African informants, often erased in published accounts, was central to the emergence of ornithological knowledge production in region. "By pushing us to focus on bodies, labors, and knowledges (of the habits and ecology of butterflies)," Hugh Raffles has stated, "relationships of this type throw questions of authorship into sharp relief."[93] Such considerations relate as well to the examination of the avian imperial archive.

By the 1880s, avian specimens and their associated documentation (e.g., notebooks, published lists, eggs) in the *Ibis*, at the British Museum, and in other museums across Britain helped naturalists such as Sclater to revise their zoological divisions of the world and map more accurately the geographic boundaries of the Mediterranean subregion, which reaffirmed and extended the European zoological boundaries into North Africa. As Jane Camerini has stated, "maps of faunal regions" such as Sclater's "served as instruments both of thought and of persuasion,"[94] illustrating visually how Britain could encounter a southern European avifauna in Morocco, Tunis, and Egypt. Such zoogeographic knowledge circulated in the reproduction of Sclater's regions in popular zoology manuals, including Cambridge professor Alfred Newton's *Manual of Zoology* published for the Society for Promoting Christian Knowledge in 1874. Robert Brown included the map on the frontispiece of *Our Earth and Its Story: A Popular Treatise on Physical Geography* (1889). In 1898, Sclater and his son, William, revised it to extend the Palearctic boundary farther south into North Africa. The map appeared in J. G. Bartholomew's *Atlas of Zoogeography* in 1911—a project funded by the Royal Geographical Society, firmly establishing zoogeography as part of the geographic tradition (figure 4).[95]

IT IS IMPORTANT to include the scientific avian specimens in the imperial archive as a means to tease out the intersection between British military culture and nature and society relationships in different parts of the British Empire. The avian imperial archive is not based in one specific institution but is dispersed widely, and though it has always been partial—a fantasy of mastery—it has acted as an ideological construction of the British epistemological worldview of order, rationality, and facts that helped to sustain the networks of empire. The collection of avian specimens by British military officers can therefore be read as multiple and overlapping systems of knowledge—through labels, lists, field notes, visual culture, and travel writing.

By linking military travel accounts, field journals, watercolors, and published articles to avian specimens, the dead bird skins examined in this book revealed how the practices of British military field ornithology—a geographic collection of birds, type localities, taxidermy, fieldnotes, and travel-writing—helped to contribute to the field of zoogeography and to materialize the

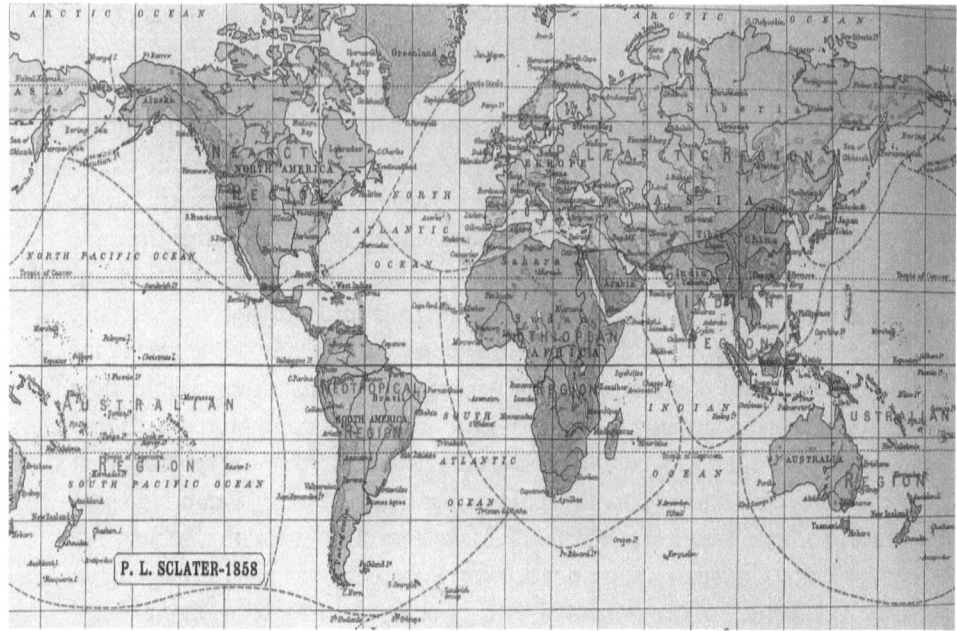

FIGURE 4  Map of Philip L. Sclater's 1858 and 1899 zoogeographic regions of the globe, in Bartholomew, *Bartholomew's Physical Atlas*, plate 2.

Mediterranean as a "semitropical" region that made visible the zoological connectivity of Europe and North Africa. As the next chapters will reveal, the extension of the zoological boundary-line into North Africa overlapped with British imperial interests in the region, transforming bird skins into geopolitical objects integral to imperial imaginings.

CHAPTER TWO

# Thomas Wright Blakiston
## *Crimean Scientific War Hero*

Avian Vignette: The Great Bustard

On 24 December 1855 in Sebastopol, British officials celebrated Christmas evening feasting on a great bustard (*Otis tarda*) during the height of the Crimean War. The winter that year was cold but not as cold as the winters of Montreal and Quebec, where some of the officers had previously served. As Royal Engineers Major George Ranken reminisced, the water on the Sebastopol harbor steamed like boiling water, much like "the St. Lawrence [River] on a very cold day."[1]

The bird, which was prepared by Mary Seacole, was said to have weighed twenty-one pounds, according to local sources.[2] Seacole, a Jamaican nurse and entrepreneur, reminisced about preparing the feast, which started in the early hours of the morning and continued until late at night. She recalled later that night dining on the bird, which had been shot for her by the Tchernaya marshes. According to Seacole, "although somewhat coarse in colour," the bustard "had a capital flavour."[3]

Known as one of the heaviest flying birds in the world, with a wingspan of up to eight feet, the great bustard would have been a welcome delight to those affected by disease, poor diets, and death during the Crimean War (figure 5). It was a common bird on the steppes of the Crimea. Officers interested in zoology were keen to seek out great bustards, since they were "seldom seen in their wild state" and had recently become extinct in Britain.[4] Bustards in Britain were often found in captivity at the zoological gardens, including the "specimens of this noble bird formerly in the possession of Lord Derby, and those at present in the gardens of the Zoological Society."[5]

Royal Artillery officer Thomas Blakiston stated, "All who were in the camp near Sebastopol during the cold weather, towards the end of the year 1855, will remember the enormous flight of bustards which passed over: they usually flew high up in the air, and, seen from the ground, they appeared quite white. Many were killed by Minie bullets fired up at them, and others when they alighted on the hills about Balaclava."[6] Blakiston also noted that the "Chersonese" was "a great resort for the bustard."[7] Lieutenant Colonel Leonard Howard Lloyd Irby shared similar details in his "Lists of Birds Observed

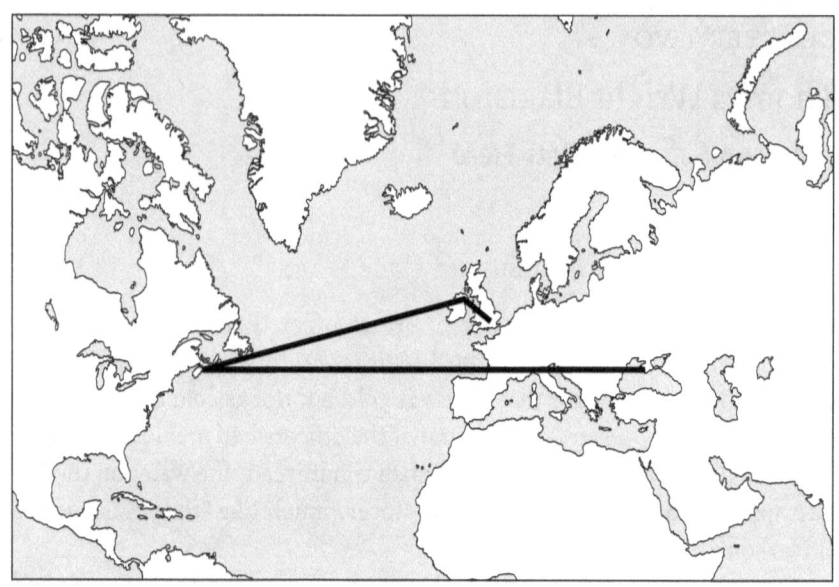

MAP 1  Route of Thomas Wright Blakiston

FIGURE 5  "The Great Bustard in Britain," from *Coloured Figures of the Birds of the British Islands*, issued by Lord Lilford (1885 and 1897). Wikimedia Commons.

in the Crimea," in which he wrote, "A flight of these birds arrived about the 5th of January, 1856, the wind at the time being N.N.E."[8] Numerous bustards "were killed and seen near Balaclava and Kamara; they are also seen about the plains near the Alma in May."[9] Great numbers of bustards remained in the region throughout the end of December and into early January 1856.

Irby would later encounter the great bustard in the vicinity of Gibraltar, where the species was plentiful in the Spanish countryside.[10] However, unlike in the Crimea, Irby could never "detect any migration of the Great Bustard" in this region.[11] He recollected how in the Crimea the birds engaged in "very large flights appeared on passage in autumn." In certain parts of the world, great bustards migrate long distances.[12] In a wild state, their diet consists mostly of grass and vegetable substances.

Great bustards, which are shy and reserved birds, were easily startled and thus difficult to approach. Officers participated in "bustard drives," which entailed driving a cart near the large birds to scare them before they were shot.[13] Irby commented on the birds' "great power of concealment," recalling a time when he attempted to catch a bustard that hid easily in the cornfield. Once the bird was exposed, Irby ran after it as it was "running along and flapping his wings"; he resorted to shooting the bird, since he could not keep up with it. He stated that he "could not have believed so large a bird could crouch so low and at the same time make such good running."

The great bustard, which many farmers considered a pest, was once part of the British countryside before the species became extinct in the 1840s, mainly due to hunting. Examples of the bird species were found at natural history museums across Britain, such as the Norwich Museum, the Linnaean Society, the University Museum at Cambridge, and the Zoological Society.[14] "A cleverly stuffed cock bustard at the Natural History Museum at South Kensington," wrote one author, showed the thirty-seven-pound bird in its nuptial display, with its head buried in the neck, and "greatly inflated." The guide to the British nesting birds of the Natural History Museum described the great bustard (Case 29) as "formerly an abundant resident on the extensive downs and plains of England, but has long since disappeared, except as an occasional visitor." Viewing a male bustard "at the height of his display" was one of the "most curious sights imaginable."[15]

Today, the global population of the great bustard is estimated at between 44,000 and 57,000 individuals. The majority (57–70 percent) of this population is found in the Iberian Peninsula, with the second-largest population center (15–25 percent) located in the southwestern Russian Federation. Great bustards are sensitive to changes in the environment due to human interference, such as agricultural intensification, power line collision, and other human-induced landscape changes.[16] Many birds found in the Crimea are protected by the Black Sea Biosphere Reserve; the most valuable settlements of the great bustard are located in the Kerch Peninsula, where the species is relatively common.[17] In the United Kingdom, there have been several attempts over the last few decades to reintroduce the species using birds from Russia and Spain, especially in southern England, such as at Salisbury Hill.[18]

> Is there any pleasure in for the first time observing a species new to you? Surely you have a peculiar feeling within you, —you eagerly wish for a specimen. . . . Suppose that you are a field ornithologist, you take the first opportunity, and although the weather is cold and windy, with snow covering the ground, you trudge off with your fowling piece to where you observed the birds.
>
> —BLAKISTON, "Birds of the Crimea"

In 1857, the Royal Artillery officer Thomas Wright Blakiston (1832–91) published a series of articles titled "Birds of the Crimea" in the *Zoologist* for a British audience interested in the natural history of distant places.[19] Writing from Camp Woolwich, England, Captain Blakiston narrated his Crimean experiences (1855–56) of "living in a single bell-tent in very hot weather, ever expecting to be on the move, and in the winter with plenty of other work, and few conveniences or books of reference."[20] Blakiston exemplified the ideal observer in the field as an officer of the Ordnance Department; officers of the Royal Artillery and Royal Engineers contributed significantly to British colonial science as they documented, listed, and classified natural history knowledge in formal and informal parts of empire. Considering that British military officers were integral to the accumulation of geographic information, natural history, and ethnographic collections in Britain and in the colonies, what role did the collection of birds from the Crimean region play in the wider cultures of the military and science in mid-nineteenth-century Britain?

This chapter investigates how the production and accumulation of scientific knowledge during the Crimean War (1853–56), a campaign waged to protect British interests in the Mediterranean from Russia, contributed to the making of a British military-scientific hero in the mid-nineteenth century. By focusing on the production of the "theater of war" in the Crimea for fieldwork and travel writing, this chapter illustrates how the natural history practices of army officials helped shape a specific heroic imaginary that embodied a rational, steadfast officer who engaged simultaneously in the natural sciences. To the general public, the war symbolized an immense human cost, with thousands of deaths suffered from neglect and disease. Blakiston's published works and collection of birds were therefore scientific trophies of war integral to the refashioning of its memory.

What follows is an analysis of what Graham Dawson terms soldier "hero-making and hagiography" and the "narrative imagining of masculinities" through the travel narratives and collecting practices of Thomas Wright Blakiston.[21] British military culture, in particular, involved a "plurality of forms" tied to national identity and an expanding empire. I begin this chapter by contextualizing the Crimean War in the British imperial imagination and the making of the modern military-scientific hero. As an institution of the state, the Royal Military Academy, Woolwich, espoused the moral benefits of the natural sciences to the esprit de corps. Next, I follow the military trajectory of Blakiston starting with his childhood experiences at Lymington, Hampshire, and his first commission overseas to Halifax, Nova Scotia. I then examine his "Birds of the Crimea" in the *Zoologist*, a monthly natural history journal founded in 1843 by the publisher Edward Newman, to uncover the gendered, class-based, and racialized representations of the European theater of war and the important distinction between "field" and "home." Finally, I trace Blakiston's avian specimens to the museum of the Royal Artillery Institution (RAI) and their meanings in terms of this site.

## The Crimean Campaign: Modern British Military-Scientific Heroes

The Crimean War, fought largely on in the Crimean Peninsula, was the first campaign to involve the European powers on European soil since the Napoleonic Wars. Britain and France declared war on Russia to preserve the security of the Ottoman Empire and, in turn, to prevent Russia from becoming a Mediterranean power and gaining access to the overland route to India.[22] Despite Britain's victory in the campaign, the experiences of the Crimean War proved detrimental to national and military morale, as British "troops were neglected and demoralized" and suffered casualties from scurvy, cholera, harsh weather, drunkenness, and "*ennui* attendant of inactivity."[23] Many accounts of incompetence and pointless deaths, including the disastrous charge of the Light Brigade, traveled quickly to Britain by media correspondents, telegraph, and photography, bringing the war zone "into the realm of technologised spectacle"[24] and affecting national sentiment at home.[25] Such negative motifs of the campaign undercut heroism with reports of failure and suffering and highlighted the weaknesses and strategic errors of the British army, all of which threatened to bring down the government and erode support for the army.[26] Colonel Arthur Hay, Ninth Marquess of Tweeddale (1824–78), expressed this alarm: "What cerebral want is in the intellectual cranium of the English which makes them so anxious to destroy all that there is of good remaining in their institutions?"[27] Colonel Hay, who was part of the aristocracy, served at Sebastopol with the Grenadier Guards and would become president of the Zoological Society of London in 1868, contributing significantly to the development of ornithology.

In order to quell public anger and reshape public opinion, military and royal officials devised strategies to shift focus onto the positive outcomes of the campaign. In 1855, Prince Albert commissioned Roger Fenton (1819–69) to produce "sanitized" views of the battlefields, which were displayed at the Royal Photographic Society's first public exhibition. Photography and art exhibitions served as important tools in the recording of the war as a historical event.[28] Queen Victoria herself commissioned the photographic firm of Joseph Cundall and Robert Howlett to photograph returning victorious soldiers to garner sympathy from the home audience.[29] One of the images from their series "Crimean Heroes and Trophies" graced the cover of the *Illustrated London News* on 12 April 1856, showing British military heroes at Woolwich carrying Byzantine paintings looted from a church in Sebastopol.[30]

Earlier that year, the *Illustrated London News* published an image of the collection of captured Russian guns and mortars brought home as trophies of war at Woolwich, demonstrating the importance of material culture and display in shaping public opinion back home in Britain.[31] These attempts overlapped with similar presentations of heroic feats in British exploration, science, and travel at the time. Dr. David Livingstone (1813–73), for example, had returned from Zambesi, Central Africa, in 1856 laden with trophies and narratives of exploration, shaping an imperial imagination of the scientific explorer-hero abroad.[32]

In order to portray a positive image at home, the mid-nineteenth-century British military adventurer embodied "the rational, prudent and calculating" and often overcame challenges with guns, compasses, diaries, and classification techniques to create order and value out of "wilderness."[33] According to Martin Burgess Green, British Protestants "became more worldly than anyone else," stressing the importance of going "abroad" where actions took place away from home.[34] The emergence of a temperate martial masculinity emphasized rational restraint, moral values, and "useful work" to control the "baser animal passions" of drunkenness, slothfulness, and luxury, and to secure British imperial interests abroad through war and peace.[35] As Major General Sir H. C. Rawlinson stated: "War, when honourably directed, being but the legitimate means of securing peace, . . . is the most honourable distinction of the soldier; but to achieve that, he must be able to support the prestige of his own reputation by the possession of real and absolute force."[36] The British military-scientific hero was "heroic" not only in battle but also during times of peace.

Many ideas about military reform and scientific pursuits emerged from the temperance movements of the 1830s at home. As one writer commented in a letter to the periodical the *United Service Magazine* in 1830, the "qualities of mind necessarily evolved and fostered in the collection of materials, are, I humbly conceive, greatly to benefit the younger branches of the service. Habits of manly thought and patient investigation will be formed, producing mental enjoyment."[37] The United Service Institution (later known as the Royal United Service Institution), a temperance society for the army and navy, established a museum "for depositing models, minerals, weapons, and specimens of Natural History, with other interesting and delightful objects, which we are daily receiving from all quarters of the globe,"[38] including Royal Navy Commander Henry Downes's "forty cases of stuffed birds and animals" after five to six years' service in Africa.[39] The building up of a collection therefore became the focal point not only for the field ornithologist but also for the moral and physical training of the military body.

While earlier reformers encouraged natural history as rational recreation, the Crimean War provided army officials with an opportunity to promote the military-scientific hero as an exemplar for future officers. In April 1859, Major General Joseph Ellison Portlock (1794–1864), Royal Engineers officer and military professor at the Royal Military Academy, Woolwich, expressed his ideologies to an audience of the United Service Institution. According to Portlock, "A nation, having resolved on a war, cannot expect to realize great events, unless by concentrating upon the object in view all its energies — physical, moral, and *intellectual*, or, in other words, its whole vital force."[40] Portlock espoused to his military audience the importance of the natural sciences in attending to "both the mind and to the body," and in sustaining "the efforts of the body politic, or of the body military."[41]

The field, in particular, provided an ideal venue to pursue "zoology and health," where military men could expose themselves to the physical exertions of the hunt and "the song of birds," exercising both mind and body.[42] When reflecting on the Crimean War, Portlock concluded: "Had the number of collectors and observers been greater" in the Crimea, "the amount of happiness would have been greater also, and that without inducing any diminution of the military zeal or efficiency of the officers themselves."[43]

British army officials looked to the fostering of scientific exploration as part of military reform and a liberal education.[44] Official bodies endorsed the intersection between travel, commerce, and public service, encouraging the pursuit of science and natural history with overlapping duties. Such organizations included the War Office, the Ordnance Survey, and the Royal Military Academy, Woolwich.[45] British military officers were active members of the Royal Society, the Royal Geographical Society (RGS), and the Geography Section of the British Association for the Advancement of Science (BAAS). Many officers of the Royal Artillery contributed significantly to and served as presidents of the Royal Society, including Sir Charles Blagden and Sir Edward Sabine, both of whom maintained personal interests in (and collections of) birds.[46]

Central to the maintenance of a temperate martial masculinity in the 1850s was a domesticated manliness, sustained by a Christian imaginative geography and an emphasis on "home," domesticity, and British women.[47] Queen Victoria herself served as a nurturing sovereign for returning "Crimean heroes" when she stood "face to face with them" at the parade of the Horse Guards in London on 18 May 1855. Her concern for wounded soldiers "brought home to the heart of the least sympathetic the ravages of war" and influenced public support for the British army in the metropolis.[48]

The establishment of rational recreation through libraries, reading rooms, and sermons also helped to promote the image of moral, temperate soldiers on the front line and during times of peace. Alicia Blackwood described a lecture on birds in Scutari delivered by the Reverend Mr. Connors as "very interesting and exceedingly well sustained, and ended with a remarkably graceful allusion to a certain sweet songster, 'whose notes were not confined to England's woods and forests, but were the solace of the sick chamber, the soother of the sorrowful, the harbinger of ease to the wounded, and the notes of a friend to the soldier.'"[49] Blackwood reiterated Connor's words: "'I need not name that bird,' said Mr. Connors, whereupon the building seemed ready to fall from the burst of applause and cheering, as every voice vociferated, 'The *Nightingale*, the *Nightingale*.'"[50]

## The Royal Artillery Tradition

From an early age, Thomas Wright Blakiston was exposed to a naval, military, and scientific life. Blakiston was born in 1832 in Lymington, Hampshire, where his family had already been involved in military campaigning in the Peninsular Wars and in Asia. His father, Major John Blakiston, the second son of Sir Matthew Blakiston (1761–1806), served with the Madras Engineers (East India Company) and the Twenty-Seventh Regiment of Foot (Enniskillens). He decided to settle in Lymingtons, a town known for servicing ships during times of war and as a port of embarkation, with fellow military and naval officers from the Napoleonic Wars.[51] The town's location along the Solent Coast, with its expansive marshes and a rich avifauna, made it an ideal spot for sporting naturalists. In 1830, Colonel Peter Hawker memorialized the region in his *Instructions to Young Sportsmen in All That Relates to Guns and Shooting*.[52]

Thomas would have been cognizant of his father's East India Company service, an account of which was published anonymously in *Twelve Years of Military Adventures in Three Quarters of the Globe* in 1829, and in *Twenty Years in Retirement* in 1835. John Blakiston discussed his strong conviction of being an Englishman while serving in India, believing that "an indulgence in field sports for want of other manly occupation" was necessary in preventing "the noble blood of England from suffering, in the same degree, as that of the high born of other countries, either from excessive refinement, or from effeminate habits."[53] Clearly his Englishness had been challenged, as he wrote, "Every man has his weak side; and one of my weak points was a dislike to be taken for an Indian."[54]

At Southsea, near Portsmouth, Thomas Blakiston attended a preparatory school. He would have seen the comings and goings of ships of the Royal Navy, army and naval personnel, and perhaps the accumulation of natural history materials from across the globe. In nearby Gosport, naval surgeon and explorer Sir John Richardson (1787–1865) housed his natural history collection at the Haslar Hospital during his seventeen years there.[55] Richardson established a library and a museum and amassed a large collection of plant and animal specimens well known to naval personnel and also to civilian naturalists, including Darwin, Hooker the elder, John Edward Gray, and Charles Lyell, as well as the younger Hooker and Thomas Henry Huxley who trained there.[56] Thomas Blakiston would, in later years, consult with Richardson prior to his expedition to the interior of British North America in the late 1850s.

Thomas Blakiston followed his father's footsteps to the Ordnance Corps as a cadet at the Royal Military Academy, Woolwich.[57] The Royal Military Academy (founded in 1741) trained gentleman cadets in surveying, cartography, mathematics, watercolors, history, and geography, all useful skills for an aspiring soldier-geographer.⁵⁸ Until 1806, cadets of the Honourable East India Company's artillery or engineers, such as John Blakiston, were educated alongside the Royal Artillery and Royal Engineers at Woolwich.⁵⁹ When Thomas Blakiston attended the academy, the institution competed with other European military schools, including the military academy in Berlin where geographer Carl Ritter (1779–1859) trained officers in the natural sciences for thirty years.⁶⁰ The Royal Military Academy taught Humboldtian science and the tracing of distributions patterns of natural phenomena across the globe as part of "truth-oriented" liberal education. This involved Euclidian geometry, which guided cartographic and mapping practices.⁶¹ By the 1860s, the Royal Military Academy included taxidermy as part of its formal training of cadets in the accumulation of geographic knowledge of colonial stations. It was "hoped that their labours" would "result in the donation to the Institution of specimens of their own preserving when opportunity is afforded them of turning their acquirement to account."⁶²

After his commission into the Royal Regiment of Artillery (usually referred to as the Royal Artillery) in December 1851, and brief service in England and Ireland, Blakiston proceeded to Halifax, Nova Scotia, in 1852. There, he encountered many British military officers who collected birds at the maritime station located on the Atlantic Ocean, driving home to him the importance of building up a collection of birds as part of a British military career. When he was stationed in Halifax, John Walter Wedderburn (1824–79) of the Forty-Second (Royal Highland) Regiment of Foot, for example, collected many avian species, including one of the soon-to-be-extinct Labrador Ducks.⁶³ The British colony of Nova Scotia was considered "wild, rugged country, covered with primeval forest and dotted with small lakes," and an ideal location for a sportsman-naturalist.⁶⁴ It was there that Blakiston established a list of the birds of Halifax with Royal Engineers officer Edward Loftus Bland and a collection of birds from the neighboring colony of New Brunswick.⁶⁵ Blakiston also networked with colonist Andrew Downs, who established an aviary and zoological garden in Halifax in 1847, and became one of his correspondents.⁶⁶ Blakiston's British North American experiences helped him to establish some scientific credibility through his collection of avian specimens, which provided the basis for his later Crimean investigations.

## The Field: "Zoology from the Seat of War"

Thomas Blakiston's first significant contribution to British ornithology involved his work during the Crimean War. As a lieutenant in the Royal Artillery, Blakiston arrived after the disastrous charge of the Light Brigade and served with allied forces from France, Sardinia, and the Ottoman Empire at the Siege of Sebastopol in 1855 to 1856. Sebastopol was an important and strategic Russian naval port in the Crimean region, which the British secured for the maintenance of the Dardanelles in the Mediterranean and Red Sea, and the security of its trade route to India. The British military network connected Crimea not only to Britain but also to other parts of the British Empire, especially the Mediterranean, with troops stationed in Turkey, Bulgaria, Romania, Greece, and Malta. The Crimean War affected other sites across the British Empire through the mobilization of military manpower. The Bosporus and Black Sea also constituted a major migratory route for numerous birds of passage, which played an active role in shaping officers' experiences of the region.

Thomas Blakiston presented a masculine, naturalist hero in the field to a British audience in the *Zoologist*, a periodical dedicated to "a popular miscellany of Natural History." He used his travel narratives to create the Crimean theater of war as a site for his military and ornithological feats during his twelve months in the region.[67] Back home in England, he was able to reflect on, polish, and order the "temporal succession"[68] of sites and avian species that completed his spatial journey in the Crimea. As Graham Dawson has noted, the pleasure of maintaining simultaneously uncertainty and familiarity required a "narrative movement that both produces excitement and suspense, and guarantees its pleasurable release and resolution."[69]

Blakiston's descriptions in "Birds of the Crimea" in 1857 heightened ideas of British, upper-middle-class masculinity involved in military prowess and codes of honor.[70] Certainly, his ornithological pursuits, like his military performances, involved continually being on the *qui vive*, a condition of heightened watchfulness or preparedness for action.[71] At camp in Sebastopol, Blakiston's attentiveness to sights and sounds prompted him to observe over his head "a flight of about fifty-six cranes steering east, but as soon as they were fairly over the camp they commenced to wheel and got gradually higher and higher till nearly lost to sight, when they bore away in about the same direction they were previously going."[72] To demonstrate his courage, he often collected birds inside the Russian lines; such escapades included "a long chase" with a red-backed shrike that he carried "for about twelve miles in a small saddle-bag, and skinned [it] the next day."[73] During the Siege of Sebastopol, Blakiston managed to obtain a species of owl of which "many were observed in the trenches," thus illustrating his gallant military service while simultaneously practicing science.[74] Narrating presence and movement in the field as a heroic naturalist provided, as Livingstone argues, "warrant . . . for the scientific stories they had to tell."[75]

Blakiston faced adversity and vulnerability in the war zone, as he suffered from "a severe fever" for two months at Scutari, which resulted in the "scantiness of notes, during summer, autumn, and the beginning of the winter."[76] Grenadier Guards officer Arthur Hay described life in the trenches as "a curious existence" in which "men suffering from every description of wounds, some dying, some just operated upon, some being dressed," and others groaning from cholera, could hear a French band "playing a polka or a waltz," as well as guns firing in the distance.[77] Similar sentiments were expressed by Royal Engineers officer George Ranken, who claimed that "disease and death are rife around me" and that "nothing makes a man feel the extreme uncertainty of life, and his entire dependence on the will of God, as much as war."[78]

To overcome hardship, Blakiston evoked the familiar through his attachment to migratory birds commonly seen in Britain, creating a vision that served to domesticate the landscape and connect Crimea to the British Empire. He encouraged naturalists to focus on the Crimea, which suggested that their observations, "if well worked up, would throw much light on the range of British birds."[79] After receiving a copy of William Yarrell's *History of British Birds* (1843), he exclaimed, "Here was my chart and compass; I could without difficulty recognize lands which were well known, and had ideas as to where to look for others."[80] He delighted in the sight of the "native" robin redbreast, which he described as a "lively little songster" that "our minds always associate with England," and "on seeing him in a distant land, wander towards that island off the coast of Europe."[81] His first meeting with the homeland bird occurred "during the armistice previous to peace," right "below the hermit's house, under the Monastery of St. George."[82] These encounters with what were perceived as British migratory birds elicited feelings of nostalgia, familiarity, and comforts of "home" in a place where many officers, including his brother, died in 1855.[83]

Blakiston used Yarrell's *History of British Birds* to order his travel and ornithological narratives in the Crimean War. Following the classification scheme of the BAAS, he divided his work into five principal avian orders, starting with "raptores [sic]," or birds of prey. Blakiston reiterated the words of Gilbert White: "'Without a system the field of Nature would be a pathless wilderness."[84] According to Mary Louise Pratt, such classification schemes helped to make European naturalists "feel part of a planetary project; a key instrument . . . in creating the 'domestic subject' of empire."[85]

While he practiced fieldwork "on a wing,"[86] Blakiston established authority in the field by advising "any one who would make full and accurate notes (for truth is the greatest point in Natural History)."[87] The collecting of specimens ensured accuracy and the production of scientific facts that could be circulated for authentication. As some contemporary critics warned, "We heartily approve of Naturalists using both their own eyes and those of their neighbours as much and as often as they please, but we as heartily disapprove of their furnishing us with long lists of critical species which they, long-sighted creatures, fancy they had seen."[88] Blakiston told his readers, when in a new country, to collect even the most common species, "for before you know where you are they may be gone never to be seen" again.[89]

The Crimean theater of war provided opportunities for many individuals to amass, present, and exchange natural history specimens. Blakiston relied mostly on a network of fellow British military officers interested in the birds of the Crimea, including Dr. William Carte, an Irish army medical officer, who amassed a significant collection of birds from the Crimea that he presented to the Royal Dublin Society.[90] When in the Crimea, Carte housed his collection at the Castle Hospital in Balaklava and received several visitors such as Sir James Alexander of the Fourteenth Regiment of Foot. Alexander, who shared an interest in birds, noted that Carte "was assisted by Lieutenant Blakiston."[91]

The European war zone revealed a variety of colonial entanglements between British and Russian, French, and indigenous peoples from the Crimean region. On the war front, the Russians maintained an interest in birds, which caused Blakiston to comment on the Russian soldiers who had constructed a number of small boxes on poles to house the house sparrow, "which may be seen everywhere in England, from the crowded streets of cities to the most remote farm-house."[92] While Blakiston did not interact with Russian soldiers in his narratives, Lieutenant Colonel Irby, Ninetieth Regiment of Foot Light Infantry (the Perthshire Volunteers), consulted with a Russian officer on the "Great White Heron (*Ardea egretta*)," who told him it "was uncommon."[93] Such an exchange illustrates how a shared interest in science superseded enemy lines during the Crimean campaign but also how military field ornithology might have served as a form of surveillance and espionage.[94]

Britain's uneasy partnership with France forced the military to work with the French army on the war front. Blakiston probably mingled with French officers in his ornithological endeavors, as he noted looking over the contents of a French soldier's haversack, which included numerous bird species, including buntings, an owl, a cuckoo, a quail, and "the little bittern (*Ardea minuta*)."⁹⁵ George Cavendish Taylor of the Ninety-Fifth (Derbyshire) Regiment of Foot met a French soldier, on his return to camp, carrying a number of grebes that he had shot in the Tchernaya with a Russian musket. Taylor, a future member of the British Ornithological Union, doubted they were good eating, but the French soldier stated that he was mistaken: "They were *poules d'eau* and *bien estimés*."⁹⁶ Taylor also saw two other French soldiers with similar species on their way back from Sebastopol, "intended for ragout."⁹⁷ Lieutenant Colonel Irby remarked on the abundance of "Purple Heron (*A. purpurea*)" in the Crimea and on how the "Frenchmen" harassed them continually so "that they had no chance of resting."⁹⁸ Many wildfowl species were found at the "French canteen," such as the "Redbreasted Merganser (*M. serrator*)" and the "Eared Grebe (*P. auritus*)."⁹⁹ In these commentaries, British military officers such as Taylor and Irby attempted to represent themselves as more enlightened than their French counterparts, who preferred pothunting to ornithology.¹⁰⁰ However, many officer-naturalists such as Blakiston contradicted themselves and killed a few birds "for the sake of being certain," but also for food (e.g., buntings) to "make a change from ration-beef and salt-pork."¹⁰¹ As Harriet Ritvo has noted, in British traveling-naturalist culture, the "protective mantle of zoological investigation guaranteed that there was no danger in going native" when eating nontraditional game or exotic birds.¹⁰²

Indigenous Tartars also were represented with similar moral rhetoric concerning birds. Crimean Tartars, especially those living around Sebastopol, had already experienced persecution from Russian authorities because of their allegiance to Istanbul rather than St. Petersburg.¹⁰³ They were reviled for their presence on white, European lands and blamed for the "bodies of thousands of Russian soldiers who lay buried in the rubble of the ruined port of Sebastopol."¹⁰⁴ Mrs. Andrew Neilson, a resident near Alma, commented on the rather "curious" ways in which Crimean Tartars killed the numerous quails in the region. According to Neilson, the Tartars "make use of a thick stick, on which the twigs are left, six or eight inches long; this they throw at the quail, giving it a circular motion, so that the bird seldom escapes being struck down either by the stick or by the twigs attached to it."¹⁰⁵ Lieutenant Colonel Irby noted similar practices in his list of Crimean birds, describing a Tartar who killed a Numidian Crane for his collection "with a stick."¹⁰⁶ Neilson's and Irby's references to the inhumane way of killing birds portrayed the Tartar people as uncivilized for a British audience back home. In the Crimean theater of war, British military officers enacted, in relation to "uncivilized" others, courage and rational restraint both in battle and in scientific fieldwork.

## Avian Scientific Trophies of "the Russian War"

British military officers such as Blakiston collected a wide variety of natural objects from the Crimean War. These specimens, destined for display at home, served as "trophies of war" and symbols of military conquest over Russian lands.[107] When HMS *Sidon* returned to Portsmouth from the Crimea in 1856, the Reverend William Henry Hawker visited the "living contents" of the cabin of naval surgeon Mr. Courtenay. Hawker observed "a magnificent eagle-owl," which served as the captain's pet, as well as some nightingales and a hobby (a small falcon). The surgeon had also pinned "a few insects too about his cabin."[108]

Many living bird species from the region were sent to the Zoological Gardens at Regent's Park in London, including "two fine birds, presented by the Commandant of Balaclava," which had been "taken near the Monastery of St George."[109] Colonel Harding presented a griffon vulture from the Crimea in September 1855, signifying an avian trophy of the Crimean War.[110] By 1846, the Zoological Gardens had opened to anyone willing to pay an admission fee, and emphasized the improvement of the more "vulgar" aspects of English society by defining appropriate ways of viewing nature.[111] Its location at Regent's Park reflected Britain's metropolitan enterprise and associated zoological riches with human privilege.[112]

As mentioned earlier, Dr. William Carte, one of Blakiston's fellow Crimean officers, displayed his collection of birds at the Museum of the Royal Society of Dublin. The museum emerged as both a "home" and a colonial museum in Ireland, for specimens donated by Anglo-Irish officers. Captain Henry of the Fourth Dragoon Guards deposited bird skins from the Crimea, including those of an eagle, a hobby (a falcon), a short-eared owl, and a little bittern.[113] William Carte's donations were described as a "large and highly interesting collection of birds from the Crimea, illustrative of the ornithology of that locality," and gained currency in his paper on the "Natural History of the Crimea," published by the Royal Dublin Society.[114] As Blakiston noted, Carte was recognized as having "the sole honour of bringing that rare bird, Richard's pipit (*Anthus Ricardi*) before the public as an inhabitant of the Crimea, a specimen of which, together with the remainder of his collection, is at the Museum of the Royal Dublin Society."[115] The museum used the "English name" of each species to instruct the general public, "especially the humbler classes," illustrating a broader imperial project of anglicizing Ireland.[116]

The majority of Blakiston's avian specimens traveled to the RAI at Woolwich alongside other military artifacts and trophies such as "howitzers, models of fortified places, Indian arms," and swords, fossils, and insects from the Crimean war zone.[117] Located on the River Thames in the county of Kent, Woolwich emerged as an important imperial site with the Royal Dockyard, the Royal Arsenal, the Royal Artillery Barracks, and, most important, the Royal Military Academy. Visitors to the grounds would have also observed the Crimean War Memorial erected in 1860 at the Royal Artillery Barracks, which commemorated artillerymen killed in the Crimean War.[118]

Henry Whitely, the curator of the RAI, managed the avian collections at the institution through "stuffing and arranging specimens of Natural History."[119] The RAI served officers of the regiment, especially those stationed at the garrison, and membership was restricted to commissioned officers. Blakiston invited Cambridge professor Alfred Newton to visit Woolwich, where he was "at liberty to look over" his collection, which "was in the hands of Mr. H. Whitely 28 Wellington Street, Woolwich."[120] By 1870, the RAI housed specimens collected by Royal Artillery officers such as Captain Graham, who donated birds from Australia; Lieutenant C. E. Souper, who sent specimens from Malacca and Singapore; Lieutenant Griffin, who added birds from Bermuda; and Lieutenant K. Gamble, who amassed eggs of various British birds.[121] Many of these officers would have been contemporaries of Blakiston.

The collections reflected a long-standing tradition of Royal Artillery officers "who had gained proficiency in the art of setting up birds, as well as preserving skins,"[122] with each case set up with the "type form" of the particular order. The collection was also arranged geographically so that visitors could compare "kindred species obtained from different parts of the globe," reflecting biogeographic interests.[123] Blakiston donated between sixty and seventy specimens of birds from the Crimea and Bulgaria to the RAI at Woolwich, which included the magnificent "Tawny Eagle (*Aquila tuevioides*, Cuv.),"[124] which, according to the *Ibis*, should be considered "a trophy of the Russian war."[125]

When he was back home at Woolwich, Blakiston authenticated his avian collections through a network of British naturalists. The authentication process included a visit by Lieutenant Colonel Irby to see Blakiston's Crimean birds, which included a purple heron, a great white heron, and a bittern shot in December 1854.[126] During a visit to the British Museum, Blakiston confirmed some of his avian specimens collected in the region; while "walking through the British Museum," he spotted a tawny eagle that resembled a specimen brought "fresh" to him by one of his friends in "the far-famed Valley of Baidar."[127] Blakiston found leading authorities, such as John Gould of the

Zoological Society of London and George Gray of the British Museum, to verify his collections. Gould examined "some specimens of the redbacked shrike (*Lanius collurio*), from the Crimea,"[128] which he considered a distinct species. Both men demonstrated "great kindness . . . in the way in which they have given [him] much valuable information,"[129] providing legitimacy to his collection of birds for an admiring scientific audience and an educated general public in Britain.

## British Military-Scientific Hero

> While our brave soldiers have been fighting our battles in the Crimea the din and glory of war have not banished from their thoughts [on] the arts of peace . . . [nor] scientific knowledge which has so suddenly been opened up to us respecting many features of the Crimean Peninsula, which so long remained a *terra incognito* to the science of Western Europe.
> —ANONYMOUS, "Crimean Snowdrop"

As we have seen, the production of the British military-scientific hero, through travel writing, bird collecting, and displaying, helped to highlight Britain's scientific achievements of the Crimean War and the connectivity of the Crimean region to the British Mediterranean through the movements of military manpower in the maintenance of the British Empire. Arising from military reforms and an expanding empire, the military-scientific imaginary promoted the pursuit of the natural sciences in the accumulation of geographic knowledge and the safeguarding of British imperial trade routes overseas.

Captain Thomas Wright Blakiston of the Royal Artillery emerged as an exemplary military hero through his ornithological fieldwork in the Crimean theater of war, encountering a variety of species, circumstances, and colonial entanglements as described in his published accounts in the *Zoologist*. According to Edward Newman, editor of the *Zoologist*, Blakiston was "an officer whose exertions in the cause of Natural History are above all praise," especially with his "admirable papers on the birds of the Crimea."[130] Blakiston's avian specimens, in particular, circulated back to the RAI and served as both trophies of war and scientific specimens. While Blakiston established the "objects' credentials"[131] in his travel narratives, his birds on display at Woolwich were transformed into avian scientific trophies of war, symbolizing the contributions of the modern British military-scientific hero within the Ordnance tradition.

Blakiston's role as a Royal Artillery officer also allowed him to gain opportunities in "imperial careering" in science and exploration: he was subsequently appointed in 1857, on the recommendation of Sir Edward Sabine, as a member of the scientific expedition for the exploration of British North America between Canada and the Rocky Mountains, under the command of John Palliser. Later, in 1859, he organized an expedition up the Yangtze River in China, making a brief visit to Yezo (Hokkaido), the northern island of Japan, in 1862. In 1870, the RAI membership could not "pass without notice the magnificent donation by Captain Blakiston, of the various birds collected by him during his connexion with the North American exploring expedition, the scientific value of which can hardly be overrated."[132] To the RAI, Blakiston was "a scientific collector who has done great service to ornithology."[133]

CHAPTER THREE

# Andrew Leith Adams
*Mediterranean Semitropicality*

Avian Vignette: The Hoopoe

The arrival of the hoopoe (*Upupa epops*) often signaled the coming of spring or the start of fall in the Mediterranean region.[1] In Malta, Andrew Leith Adams described how "the weird-like form of the hoopoe may constantly be seen drifting before a south wind in spring, or hastening southward in August, seldom in flocks."[2] Writing in 1865, Adams noted that the hoopoe usually traveled individually and alighted on the wind currents of the seasonal sirocco, making "its appearance in Malta about the 7th of April, and again in the beginning of September."[3]

The unique-looking bird, with its feathered tuft, pinkish-brown body, and striking black and white wings, is prevalent across Europe, Asia, North Africa, and sub-Saharan Africa and connects various landscapes and habitats from Britain to tropical Africa (figure 6). The majority of hoopoes in Europe, such as the ones observed in Malta and Gibraltar, migrate to the tropics in Africa during winter.[4] "This handsome bird," the editors of the *Guide to the Gallery of Birds in the Department of Zoology, British Museum* (1919) described, "is a spring visitor to the southern and eastern parts of England, where, if unmolested, it would breed regularly."[5] As hoopoes were regularly persecuted on their arrival in Britain, "very few pairs survive[d] and [were] allowed to rear their young in peace."[6]

Maurice Drummond Hay, who had served in Corfu and Malta with the Forty-Second Regiment (Black Watch) in the 1840s, believed the Mediterranean seaboard was an ideal site to study the migratory patterns of the hoopoe, "especially in the spring passage from North Africa in March and April; and at Malta, so common were they at that season."[7] Drummond Hay first heard the bird in the olive groves of Corfu, although he did not know that they preferred open fields instead. The bird's name is based on its distinct call note, "sounding like a 'hoop hoop' two or three times repeated in a low guttural tone."[8] Drummond Hay had sent a specimen from Tangier to the Zoological Society and noted they were seen regularly in northern Africa among "dung-hills."[9]

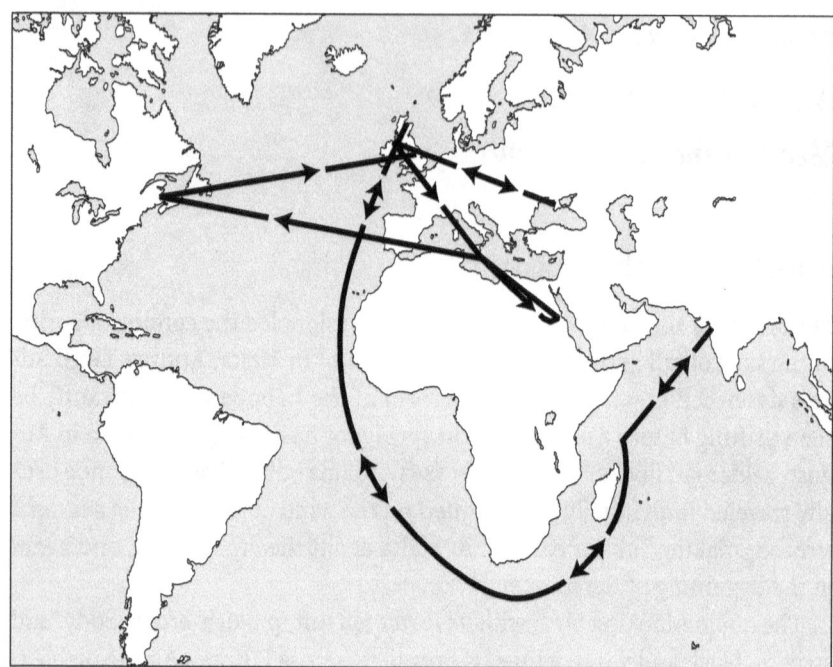

MAP 2  Routes of Andrew Leith Adams

According to Leonard Howard Lloyd Irby, the hoopoe, which he encountered around Seville in Spain, when he was serving at Gibraltar, had a call that sounded more like "hood, hood."[10] Although hoopoes did not breed at Gibraltar, they arrived in the spring on "the 17th of February, 1870; 18th of February, 1871; 16th of February, 1872."[11] Most hoopoes crossed Gibraltar in March, Irby observed, "whence their local name *Gallo da Marzo*, Marchcock."[12] Captain C. W. Watkins in 1856 noted that many of these birds were "found for sale in the Gibraltar Market" during this month.[13]

Military officers often viewed the breeding hoopoe in the vicinity of Gibraltar. Irby observed, for example, these birds nesting around Casa Vieja, Andalucía, Spain. He recorded how many lay eggs around 1 May, in the holes of trees.[14] As ornithologists have observed, hoopoes are monogamous birds on a seasonal basis and are territorial, with the male calling frequently to advertise his ownership of the territory.[15] As Irby noted, "The nest is formed of bents, and lined with soft materials; it is built in the hollow of a tree, and is said to be extremely fetid. The eggs are four in number, bluish white, spotted with pale brown."[16]

FIGURE 6 Sketch of a hoopoe by John Gerrard Keulemans, in Salvin and Hartert, *Catalogue of the Birds in the British Museum*.

Officers stationed in the Mediterranean crossed paths with the hoopoe in various parts of the region. Hoopoes were observed in Turkey during the Crimean War, where George Cavendish Taylor of the Ninety-Fifth (Derbyshire) Regiment of Foot commented, "There is plenty of ground and opportunity in Turkey for a sporting ornithologist." There, hoopoes were abundant during the spring migration, when he observed thirteen birds in one day while at camp near a Russian burial ground. These birds were accompanied by other migrant birds, such as rollers.[17] Colonel Williamson of the Ninety-Second Regiment met with "vast numbers, near Ceuta, in Africa, opposite to Gibraltar, during the whole year (ca. 1820s–30s)."[18] Adams viewed several hoopoes on his voyage down the Nile River, where he observed many flying with other European migrant birds when the birds passed his vessel. He noted that the birds could be traced back to the ancient Egyptians, based on a painting of the hoopoe he observed on tombs of Beni Hassan in Egypt.[19]

Using notes from the late M. F. Favier, Irby included details about the bird in North Africa, near Tangier. According to Favier, hoopoes were "seen in great quantities near Tangier on passage, crossing to Europe during February, March, and April, returning, to retire altogether for the winter, in August, September, and October" in the 1840s through 1860s.[20] The French consul-naturalist documented the importance of the species to the Jewish and Islamic cultures in North Africa, both of which believed "that the heart and feathers of the Hoopoe are charms against the machinations of evil spirits."[21] In some cultures, the hoopoe was believed to possess marvelous medicinal qualities, and certain parts of the hoopoe possessed magical powers, including the heart, blood, eye, head, tongue, wings, and feathers. Hoopoes were known as water finders in ancient times.[22]

Hoopoes often made it as far as the British Isles, where "scarcely a year passes without some specimens being obtained, sometimes in the south, and at others almost in the extreme north."[23] Black Watch officer Maurice Drummond Hay produced a specimen for display for the Perthshire Society of Natural History's museum collection, which he obtained by H. Wedderburn on the south bank of the Tay at Birkhill, Scotland. According to Drummond Hay, hoopoes were a rare occurrence in Scotland despite being "a veritable bird of passage."[24]

Today, the hoopoe is on the International Union for Conservation of Nature (IUCN) Red List of Threatened Species, as defined by BirdLife International.[25] It is seen as a valuable species to many, considering its diet consists of many insects that humans view as pests; the species is afforded legal protection in many countries.[26] In Malta, the hoopoe has been protected since 1980 and is one of the species highlighted in BirdLife Malta's bird protection campaign. In the United Kingdom, the hoopoe is often viewed as an exotic bird, although with climate change it might become a frequent visitor in the coming years.[27]

> Malta is so little known to English readers, except in its qualities of a fortress and a coaling-station, and so generally considered as little more than a huge rock, that ornithology in connexion with it seems almost paradoxical. Burnt up and barren under the African sun of its summer, with the rains of winter it rises, like a phoenix, from its ashes to verdure and life.
>
> —WRIGHT, "List of the Birds"

"When I first set foot on the Maltese islands, June 22, 1860," reflected British military surgeon Andrew Leith Adams (1827–82) in his *Notes of a Natu-*

*ralist in the Nile Valley and Malta* (1870), "an impression came over me that I could not have selected a more uninviting and uninteresting locality for the study of the natural sciences."[28] For Adams, who spent his childhood in northwest Scotland, Malta was "bare, weather-beaten, rocky, and sterile to a degree, no woods, and scarcely a tree to be seen anywhere."[29] He attributed the lack of verdure to the "semitropical sun and the dreaded sirocco" that dampened his "ardour in the pursuits of natural objects."[30] His first impressions, however, quickly disappeared once he experienced more temperate weather conditions and witnessed the migration of birds visiting the islands en route from Europe to Africa. After seven years' residence in Malta, Adams "found ample occupation in making collections and noting the names and numerical prevalence of the various birds of passage" and was able "to estimate the relative proportions" of each species "with considerable certainty."[31]

Adams's life as a military surgeon with the Twenty-Second Regiment of Foot allowed him access to places that were off limits to the regular traveler and naturalist. His movements through different imperial sites created opportunities for comparison of birds, peoples, and landscapes; helped him build up a vast collection of bird specimens; and prompted him to publish a series of books based on his military naturalist experiences in India, Malta, Egypt, Nubia, and New Brunswick. Adams, who had a particular fondness for birds and field ornithology, extolled to his "confrères" in the army and navy the virtues of "physical studies as remedies for idleness during the many leisure hours spent in often less profitable undertakings, for Nature's field is broad and inviting."[32] According to Adams, natural history fieldwork invigorated military discipline and revitalized "the mental powers," and thus supplied "materials for the grandest ultimate truths."[33] He espoused ornithology as "a branch of liberal education" and a means for "mental improvement" rather than for "utility, as applied to the physical wants and material interests of mankind."[34]

This chapter builds on notions of a temperate martial masculinity introduced in chapter 2 by focusing on the transimperial career of military surgeon Andrew Leith Adams and his contributions to the maintenance of military and racial fitness through field ornithology and the British army medical tradition.[35] The complex lineage of the term "temperate" can be traced back to Aristotelian ideas of a temperate zone. For Aristotle, this climatic zone, which encompassed Europe and the Mediterranean region, was viewed as the only region capable of fostering habitation and civilization. However, Aristotle also extended his notion to a temperate embodiment tied to good mental health and a sound mind as a means to prevent corruption from vices.[36]

More recently, the geographer James Duncan has used the term to specify both a geographic region and a gendered and racialized embodiment or "temperate masculinity" within the imperial project of coffee plantations in mid-nineteenth-century Ceylon. Drawing from the work of David N. Livingstone on the "morality of climate," the "temperate" involved the physical conditions of higher altitudes of Ceylon, and an imagined place for the re-enactment of a highland Europe in relation to the Singhalese heat and jungle. Central to this understanding was the designation of "tropical nature" as an agent of disease, crop failure, and slothfulness.[37] A moral or temperate masculine embodiment therefore involved codes of conduct of proper behavior or "moral hygiene" and a European masculine sense of self-control, safeguarding the survival of British middle-class, white coffee planters in colonial Ceylon.[38]

In her discussion of British military culture, Sonya Rose has used the concept of "temperate masculinity" to analyze the connection between tempered British masculinity and nationalism during the Great War (1939–45). While set in a different time period, the formulation of the British military hero emphasized "reason" as a quality associated with both masculinity and civic virtue. Rose traced this form of masculinity to the early Victorian years and the manly code of behavior taught to young boys at public schools. A temperate masculinity focused on fair play, tolerance, and kindliness as important attributes for the "modern" idea of "character."[39]

In this chapter, I extend Duncan's and Rose's notions of temperate masculinity to consider the impact of a career in "empire" on the formation of gendered and racialized masculinity in mid-nineteenth-century British military culture, and its relation to ornithological practice. By the 1860s, the British government was investing in health reforms dedicated to the improvement of the army in the tropics to cut military costs.[40] These reforms led to the prevention of lengthy stays in tropical stations by the 1870s. The study of the effects of climate on soldiers could help in "fixing the duration of the sojourn of foreign troops at certain stations, so as to render them effective in war."[41] Reforms involved disciplinary spatial practices that attempted to control the lives of military men in the "name of national morality and martial strength."[42] These included shaping their "recreational geographies."[43] Adams published several articles on army recruiting and the effects of climate on the military mind and body, and he believed in, "as a rule, the possessors of a *mens sana in corpore sano*."[44] Following the example of the Scottish explorer-hero David Livingstone, who "weathered the storm, 'the hardships of sun and soil' must be overcome."[45]

What follows is an examination of Adams's "temperate approach" to military life and to field ornithology. I trace the influences of the Scottish civic science and temperance movements, as well as the natural history traditions of the Army Medical Department.[46] I then follow Adams's military career to India, where he encountered tropical and temperate zones and consider their impact on the racial degeneration of the white, European body. To Adams, natural history fieldwork as a hygienic practice, or what Livingstone has termed "regimens of bodily management," had the effect of preventing the harmful effects of tropicality on both the body and the mind.[47] Finally, I focus on the Mediterranean station of Malta and on how Adams helped to materialize the Mediterranean region as a "semitropical" zoological subregion for the physical and cultural acclimatization of white, transient British officers to and from India. In doing so, he made "visible in new ways" the connectivity of North Africa to Europe through his contributions to ornithology and geology. These ideas became solidified as he completed his military service in New Brunswick and experienced New World wilderness and a northern climate.

The British Army Medical Tradition

Born in 1827, Andrew Leith Adams grew up in the small village of Banchory-Ternan in Aberdeenshire, northeast Scotland, where his father, Dr. Francis Adams (1796–1861), established a medical practice, and raised his children following the death of his wife, Elspeth Shaw.[48] Adams's father encouraged his sons to study the local natural history of the "beloved haunts of his nativity" along the "Banks of the Dee, and among the Grampian Mountains of Scotland."[49] Andrew and his father collected many examples of the local avifauna, and preserved them in the family "Museum of Natural Curiosities" in Banchory.[50] There, Andrew encountered the "Golden Eagle in his native place, which flew across the Grampians with its legs entangled in a fox-trap" and the "Snowy Owl" in the woods near "Blackhall House."[51] As a Classical scholar, Francis Adams specialized in the works of Hippocrates, such as *Airs, Waters, Places*, which laid the foundation for studying the effects of the physical environment on living organisms over an extended period of time.[52]

IN THE 1840S, Andrew Adams attended medical school at Marischal College, University of Aberdeen, which emerged as an important site for the

British military medical tradition, as well as a center for Scottish "civic nationalism."[53] The college offered courses in geography, and many of its members were involved in the British antislavery movement and missionary work, as well as the military and exploration, including James Augustus Grant (1827–92), an East India Company officer and Scottish explorer of eastern equatorial Africa.[54] One of its wealthiest patrons, Dr. Robert Wilson (1787–1871) donated his archaeological specimens collected on his travels to the Marischal Museum (est. 1786) and offered a traveling scholarship to students interested in exploring Asia and Africa.[55]

The college was linked to the Scottish temperance movement that imagined progress through respectable leisure activities, mental improvement, and "social control."[56] The temperance tradition in the west of Scotland started as an anti-Catholic reaction to the Irish immigrants who settled in Aberdeen following the 1830s potato famines in Ireland. Scottish reformers blamed Catholic Irish immigrants for many urban social problems in Scotland, including alcoholism.[57] The Reverend George Wisely, a proponent of Free Church missionary endeavor, attended Marischal College and would later bring his temperate reforms to Malta.[58]

Marischal College also was home to William MacGillivray, a well-known Scottish naturalist and ornithologist, who became a professor of natural history in 1841 and published numerous works such as *A Manual of British Ornithology* (1840–42) and *A History of British Birds, Indigenous and Migratory, in Five Volumes* (1837–52). MacGillivray gained prominence for assisting Robert Jameson, the Regius Professor of Natural History at the University of Edinburgh and for acting as curator of the museum of the Royal College of Surgeons of Edinburgh. He condemned "cabinet naturalists" and taught his students the value of collecting specimens in the field, influencing many future naturalists, including Adams. More specifically, MacGillivray encouraged Adams to devote himself "to prolonged studies in natural history" when a surgeon in the army.[59]

Adams entered the British Army Medical Department as assistant surgeon with the Ninety-Fourth Regiment of Foot in 1848, a department known to promote the synergies between army surgeon life and works in natural history. In *A Catalogue of the Collection of Mammalia and Birds in the Museum of the Army Medical Department at Fort Pitt, Chatham* (1838), the editor highlighted the contributions of Sir James McGrigor, director-general of the Army Medical Service (1815–51), and encouraged medical officers in all parts of the globe to pursue natural history.[60] McGrigor, another military surgeon from Marischal College, served in Egypt and the Napoleonic Wars with the

East India Company and also created the Royal Army Medical Corps.[61] He also established the Fort Pitt Museum of Natural History at Chatham, which housed a "distinct collection" of zoology amounting to 9,386 specimens by 1838, including birds from India, Bermuda, British North America, South Africa, New South Wales, and Trinidad.[62] The museum was well known to early-nineteenth-century naturalists such as John Richardson and William Swainson, who used the collection for their ornithological works.[63] Adams would later choose to donate his specimens from India to this museum.[64]

## Tracing the Contours of Tropicality and the Temperate

Adams's notions of tropicality and the temperate were shaped by a transient career in different climatic regions of the British Empire. As David Livingstone has noted, the moral discourse of climate and empire was coconstituted by and tied intricately to "white labour" in the tropics.[65] In 1849, Adams set sail for India around Cape of Good Hope to serve in the Second Anglo-Sikh War (1848–49).[66] While in India, Adams traced the contours of tropicality and its effects on the British military body. Medical doctrine on acclimatization in the early nineteenth century centered on the "seasoning" of European troops as an adaptation strategy for service in the tropics.[67] By 1870, however, lengthy stays in the tropics were avoided at all costs, as doctors were documenting the gradual deterioration of the European body in tropical regions.[68] Such studies garnered attention, since fostering military fitness among the troops was central to maintaining Britain's imperial links to the rest of its territories and reducing the cost of empire.

British India became synonymous with what David Arnold has described as the "restless movement" of territorial boundaries and the mobilization of military personnel, as the East India Company engaged actively in territorial acquisition in the first half of the nineteenth century.[69] Adams's mobility with his regiments, first with the Ninety-Fourth and then with the Twenty-Second, was reflected in the ways in which he cataloged, described, and listed the birds in northern India during his postings to Dagshai, Rawalpindi, and Peshawar.[70] In a sense, Adams's collection of birds in India enacted an imperial performance of territorial expansion, especially in places such as the interior of Ladakh, which was unknown to Europeans. His naturalist and imperial legacy would be commemorated in the black-winged snowfinch (*Montifringilla adamsi*) at the British Museum in London, denoting territoriality defined by the Queen's Army as opposed to the East India Company.[71] According to Peter Stanley, even though both armies served together at war, there existed many tensions between the two cultures, resulting in the incorporation of the East India Company army into the Queen's Army following the "Indian Mutiny" in 1858.[72]

In India, Adams embodied the enlightened European naturalist by referring to Alexander van Humboldt's *Cosmos* and appreciating "authentically" the "beauties of nature."[73] Adams believed that journeying "over the torrid zone" to experience "the luxuriance and diversity of vegetation, not only on the cultivated sea-coasts, but on the declivities of the snow-covered Andes, the Himalayas or the Nilgery mountains of Mysore" allowed him to appreciate fully the aesthetics of the natural world.[74] In his writings, the emergence of the "torrid zone" or "tropicality" signified a place of enchantment and fecundity, which rendered distant regions such as India accessible to the British imagination; these regions also opened up the possibility of resource extraction for the empire.[75] Adams's romantic tropes also worked to domesticate foreign places. As he wrote: "Few Englishmen could sit on the grassy banks, and witness the rare mountain beauty of Arabel without a feeling that did Cashmere belong to England, there is no spot among all its lovely scenery better suited for a pic-nic."[76]

Tropicality, however, also engendered the unfamiliar and posed a threat to the white, European body. Adams included a section titled "Deterioration of Race" in his book on the natural history of Malta. "The white man may live under the equator," he wrote "but his race will deteriorate. . . . unless by constant infusion of fresh blood, there will, in a few generations, take place a deterioration of race so marked that time seems only requisite to bring about entire extinction."[77] Adams noted the effects of tropical climes on the English "race" when describing an encounter with an Anglo-Indian in India. The man "was an example of a race of Englishmen born and brought up in India without the shadow of an idea of anything beyond Hindostan and its European society, and even the smallest portion of thought on these points, for in his manners he had most in common with the native, whose language he spoke more fluently than his own."[78]

For Adams, the pursuit of natural history served to minimize the perceived racial and masculine degeneration of officers serving abroad in tropical climates such as in India.[79] Adams believed that the adherence to these "simplest of hygienic rules" could negate "the so-called insalubrity of the climate" and therefore maintain an officer's constitution in the tropics.[80] He viewed natural history as a "requisite for the army surgeon" and an activity that prevented the harmful influence of military life "in all climes" and in "varying conditions."[81] The avian scientific specimen provided British military officers with tangible proof of rational recreation and self-improvement in overseas colonies in order to prevent the perceived degeneration of "mental culture" in tropical environments. The forming of a collection of birds helped to maintain "temperate masculinities" through respectability, rational thought, and moral recreation, which formed part of the process of disembodied rationalism of Western discourse.[82] Ornithology therefore became one of the key strategies to limit the perceived effect of colonial environments on military bodies and, in turn, on military colonial knowledge production.

Adams connected with his Scottish temperate homeland through northern India's avifauna, which he viewed as "a good many denizens of the air whose brethren he had been familiar with in his infant days, and which accordingly gave rise to the most pleasing remembrances of antecedent sights and scenes of delight."[83] These birds awakened in "his mind the recollection of his *'natale solum,'*[84] from which he had been estranged in early life."[85] In northern India, the woodcock, snipe, and plover "on the Ghants, or Nilgiris" helped him recall the "Woods of Blackhall, and the Loch of Leys" of Aberdeenshire, while the heron on "the waters of the Indus, or the rivers of the Punjaub," brought back "his infant days" on "the waters of the Feugh or the Dye."[86] The "Indian robin, so generally distributed over most parts of Hindostan," differed in appearance from the European robin or robin redbreast but exhibited similar habits such as "jerking its tail as it hops along."[87] Adams exclaimed: "How often have associations of home been brought to mind by seeing this pretty little warbler pursuing its gambols before the door of an Eastern bungalow!"[88] Spatially linked to "home," and to the homely bungalow, the foreign bird's familiar movements helped to domesticate their presence in colonial sites, and to discipline the senses in the more temperate regions of British India.

More important, Adams's work on mapping the range of birdlife in India helped to demarcate the more temperate regions of India at higher elevations, especially at the different hill stations. These locations became important sites for the biological and ideological reproduction of British life or, as Dane Kennedy has stated, as "the nurseries for the ruling class."[89] According to James Bird, a physician with the General Bombay Army, hill stations were "suitable for the healthy residence and vigorous existence of the European race," and more specifically "congenial to the feelings and health of Englishmen."[90] Adams's literary mappings included "the pool near the village of Sehwan, which during the cold months is covered with wild-fowl; here we procured specimens of the shoveller, castaneous and tufted ducks, the Gargany teal, "and here I met for the first time the spotted-billed duck."[91]

*Andrew Leith Adams* 51

Adams also mentioned Kandala, where "the traveller is surrounded by a varied fauna and flora. What finer sight than that which greets him at day-dawn on some cool November morning."[92] These mappings served to delineate the climatic boundaries for the maintenance of British, white, racial identities in India, and influenced his ideas of semi-tropicality in the Mediterranean region.

## Mediterranean "Half-Way House"

After a brief service in the Crimea as a volunteer, and then a short return to Britain, where he rejoined his regiment at Manchester and Dublin, Adams moved on to Malta, where during his leisure time he spent six years "studying the ornithology of the region, more especially the migratory birds which pass and repass annually on their ways to Europe and Africa."[93] By the 1860s, Malta had emerged as an important site in the Mediterranean for the maintenance of the British "empire route" to India, and a strategic location for the efficient rotation of troops across the empire.[94] With increasing anxieties over military fitness and racial degeneration in tropical climates, British military officials such as Adams constructed the Mediterranean as a semitropical zone. This meant that the seasonal variations of climate, the symbolic connections with homeland, and the racialization of southern Europeans helped to maintain British military identities and to naturalize their presence in the Mediterranean region. The production of semitropicality therefore helped to shape a "moral-political landscape" for the cultural acclimatization of the "homeward and outward" British military traveler to and from India.[95]

Winds played an important role in the designation of the British Mediterranean as a semitropical place and in the "embodied performance" of enacting the semitropics.[96] In Malta, military men encountered the sirocco, a southeast wind emanating "from the African deserts," which reportedly "depresse[d] the energies of the mind, and produce[d] lethargy, and low spirits."[97] In a report on the "meteorological phenomena in connexion with cholera and other diseases," based on his recordings of daily temperatures at the governor's house in Malta, Adams believed the influence of the southeast wind on people to be "a good deal exaggerated by writers."[98] The "disagreeable effects" were mostly experienced in September, when the wind "occasioned a feeling of lassitude and inertness, accompanied by increased perspiration, headache, and often irritable boils."[99] The stifling impact of the summer months eventually gave way to more temperate conditions, when, according to Adams, "studying nature in such a climate . . . is certainly most delightful. The temperature, continually mild, renders exertion pleasant, and one never feels overdone by walking fast, or the necessity of doing so in order to keep up the animal heat."[100]

To help maintain temperate embodiments, the Malta Garrison Library[101] in Valletta housed a large collection of books devoted to geography, travel, exploration, and natural history donated by Governor William Reid (1791–1858), who served as governor of Malta from 1851 to 1858.[102] As a Royal Engineers officer, Reid committed himself to the natural sciences, especially meteorology.[103] The formation of the library overlapped with other temperance movements by Scottish reformers on the islands, which included the establishment of the Free Church of Scotland.[104] The Garrison Library held a copy of Andrew Leith Adams's *Wanderings of a Naturalist in India* (1867) and other naturalist works that included instructions on how to conduct taxidermy and amass a collection of natural history objects.[105] The library, restricted to the use of the officer class, also housed natural history curiosities donated by officer gentlemen. As Douglas Peers has noted, although the culture and rhetoric of the officer class prized a middle-class virtue of respectability, few officers wished to transform the other ranks through these moral reforms: "plebeian culture" suited military interests.[106]

Britain's proximity to the Mediterranean allowed for a regular flow of British travelers en route to Egypt and Asia, including Alfred Russel Wallace, who stopped in Malta from the Malay Archipelago and purchased a hoopoe in the market in March 1862.[107] Military men often quartered at Malta during their travels from India to Britain. Major General William Denison, for example, spent time at the Mediterranean station and inspected the fortifications.[108] Malta had already experienced a long tradition of travel as part of the Grand Tour during the Order of Saint John or Knights of Malta (1530–1798), and briefly during French rule (1798–1800).[109] British travelers visited the Mediterranean islands as part of health tourism until the 1850s.[110] English writer William Makepeace Thackeray likened Malta in November to England in May.[111]

More important, such a tradition of travel facilitated an increasing presence in the Mediterranean of British middle-class women who could mediate imperial, moral masculinity and the "regulation of manhood."[112] As Linda Colley and Dane Kennedy have noted, the running of the state depended on the contributions of middle-class women to men's lives and the education of children in the homeland, and especially in "the colonial realm" for the maintenance of power, identity, and British values.[113] The *Malta Times* featured a special column that listed all of the "Winter visitors" from Britain, which included Jemima Blackburn (née Wedderburn), one of Britain's leading bird painters, who visited Malta on her way to Egypt and illustrated "Denizens of ancient Malta" in Adams's book.[114] Military wives, as well, could accompany their husbands abroad. Adams's wife, Bertha Grundy Adams, followed him to Malta and played an active role during the cholera epidemic, chronicling her husband's regimental activities.[115] By summertime, however, many winter visitors would make "off like rats from a sinking ship" when "the gaieties of the Malta season" were over.[116]

British winter visitors also included "the various birds of passage which make the [Maltese] islands their half-way house."[117] While Andrew Leith Adams engaged in geologic investigations "to inquire into the capabilities of the islands from a natural-history point of view,"[118] birds remained central to his naturalist explorations, and especially the migratory birds that traveled annually to Europe and Africa. Adams reveled in the return of Britain's migratory birds, noting that "the annual migratory visitors and accidental arrivals amount to no less than 240 species."[119] When the hot days of summer returned, however, the "gay songsters" that crowded "every available bush and tree and field" fly back to "more northern climes."[120]

Adams claimed that "nowhere are the feathered tribes more persecuted than in Malta," where "one-half of the migratory species are captured or shot, and of all days, on Sunday the greatest carnage is perpetrated, so that on the following day the poulterers' shelves are stocked with all manner of birds, great and small."[121] Local Maltese epicure included "the pretty little scops-eared owl and goat-sucker."[122] Moral assertions about the pothunting practices of the Maltese, which attributed "beastly alterity" to southern Europeans, fixed "a racial hierarchy and an assertion of essential differences between natives and their colonizers."[123]

Throughout the nineteenth century, British officials debated the racial identity of the Maltese. According to British colonist and naturalist Charles A. Wright, "The Arab houses, language, and origin of the inhabitants indicate, despite Acts of Parliament and a European fauna, Malta's alliance with Africa and the East."[124] Adams described the "kind hearted rustics of Malta" when conducting fieldwork in the countryside; he reflected: "I can picture the brown-skinned and wiry son of toil, with his nightcap-like headdress and tight fitting garments."[125] To Adams, they were often "half-confused" and called him "Inglese" in their "mongrel Arabic."[126] As Sandra Scicluna and Paul Knepper have noted, the Maltese became progressively "whiter" when Britain supported self-government for Malta in the early twentieth century.[127]

## European Zoological Connectivity with North Africa

The ambiguity of the geographic location of Malta emerged as British officials contemplated the island's connectivity to North Africa. Many British naturalists did not know where to position Malta geographically in the Mediterranean region. According to Robert Montgomery Martin, the island of Malta "was formerly placed by all geographers in Africa, but was declared to be in Europe, as regards to the service of our soldiery, by a British act of parliament."[128] Oxford University professor Hugh Edwin Strickland wrote in 1850: "I hardly know whether the occurrence of a new or unrecorded species of bird at Malta is to be regarded as forming an addition to the European fauna, because geographers are I believe not yet agreed as to whether Malta belongs to Europe or to Africa."[129]

Early French military investigations in Egypt, Algeria, and the Morea in the 1830s had already mapped the Mediterranean Sea as a transition zone for different species between the three continents of Europe, Asia, and Africa.[130] In the 1840s, the British Admiralty commissioned a number of survey ships to produce hydrographic charts and maps of the Mediterranean to make it a safer place to navigate.[131] Captain Thomas Graves on board HMS *Beacon* in April 1841 observed the spring bird migration "when many species of birds which make Europe their abode only in the more genial seasons, were, after having passed the winter in Africa, crossing the Mediterranean to their summer quarters."[132]

Adams recorded and collected the various migratory birds that traveled to and from Europe to Africa. He concluded: "From an ornithological point of view," the Maltese islands served as "winter homes of European birds of passage," which extended the European boundaries into North Africa. These birds, especially the ones in Tunis, Algeria, and Morocco, "although differing in physical features and, in some respects, in climate, are, strictly speaking, but an extension of Europe, for their flora and fauna are European."[133] Adams posited that "it is only when the traveller crosses the Sahara, with its salt lakes and moving clouds of sand, and gains the region of verdure beyond, that he enters on a new zoological and botanical province."[134] Adams's notion of zoological provinces followed a single origin of species (monogenism) rather than multiple origins (polygenism) or creation sites.

Adams's ornithological knowledge of Malta was based on a network of British naturalists—Charles Augustus Wright, John Gould, Charles Bree, and Sir William Jardine—and the collection of avian specimens acquired in the Mediterranean region. He recited the list of Charles Wright in his book on the natural history of Malta, which could only be viewed at the Malta Garrison Library.[135] Wright's list built on the work of Maltese naturalist Signor Schembri in 1843, and consisted of 253 species in 1864, with about 10 to 12 species identified as resident while the others made Malta a "resting-place" for their "periodical migrations across the Mediterranean."[136] In documenting these movements, Adams's networked ornithological knowledge helped to dispel early theories on bird migration such as the idea that "certain birds spent the cold months at the bottom of lakes."[137]

Adams also engaged in geologic investigations of the connectivity of Europe to Africa.[138] As Charles Wright described it, Adams pursued "a geo-ornithological voyage of discovery" around the "great fault at Naxar."[139] Influenced by the works of Charles Lyell and Captain Spratt, Adams described the geology of the islands and discovered fossils or remnants "of a by-gone age—the pigmy elephants, the hippopotamus, the great extinct swan and fresh-water turtle, and the great dormouse."[140] These discoveries were published in popular works on the Mediterranean region such as *Handbook to the Mediterranean* (1881) by Lieutenant Colonel Robert Lambert Playfair, who was consul general of Algeria. According to Playfair, Malta "must be regarded as fragments upheaved of the sea-bottom which connected Europe with Africa."[141]

British imperial interests in North Africa centered on informal empire, which, as John Darwin has described, encompassed the links fostered by trade, investments, or diplomacy in order to draw new regions into the world-capitalist-system and, more specifically, Britain.[142] European foreign policy in the region focused on maintaining neutrality in Morocco, Tunis, and Egypt as a means of securing a stronghold over the Mediterranean trade route to Asia, especially with the opening of the Suez Canal in 1869. For example, both France and Britain negotiated for the Anglo-French institution of Dual Control (1879–82) in Egypt, which brought an end to French domination in the territory.[143]

When on leave in March 1858, Adams took an excursion to Tunis from Malta on board the HMS *Wanderer*.[144] As a protectorate in the 1850s through 1870s, Tunis remained important in Britain's informal empire in North Africa, with many Maltese living in the territory and fostering trade and commerce networks between Malta and North Africa.[145] Britain attempted to consolidate ties with the Ottoman Empire as a means of preventing France from extending territorial interests outward from Algeria.[146] Britain relied heavily on Tunis for bullocks, sheep, fruit, and vegetables to be shipped to British military garrisons in the Mediterranean; in return, Britain supplied

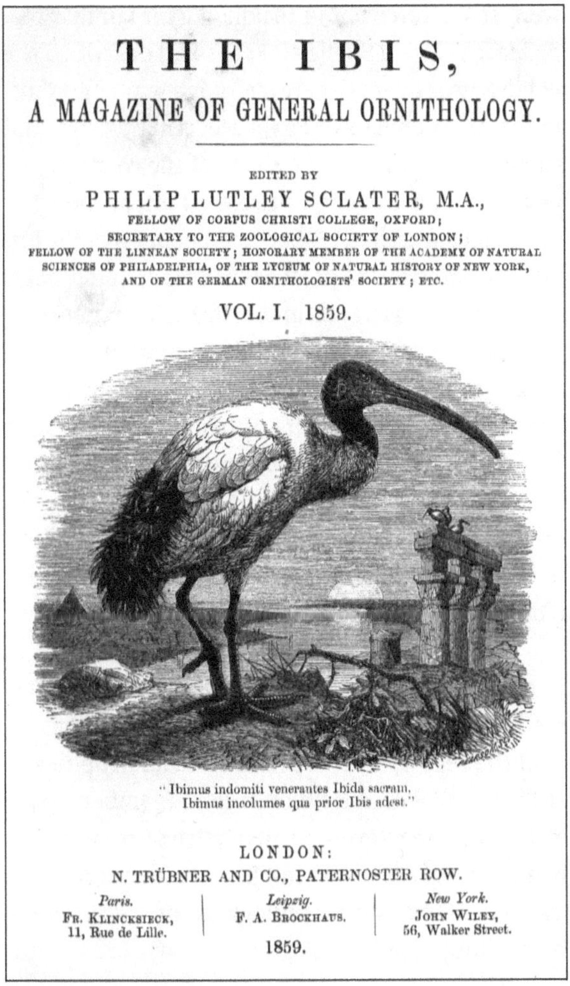

THE IBIS,

A MAGAZINE OF GENERAL ORNITHOLOGY.

EDITED BY

PHILIP LUTLEY SCLATER, M.A.,

FELLOW OF CORPUS CHRISTI COLLEGE, OXFORD;
SECRETARY TO THE ZOOLOGICAL SOCIETY OF LONDON;
FELLOW OF THE LINNEAN SOCIETY; HONORARY MEMBER OF THE ACADEMY OF NATURAL SCIENCES OF PHILADELPHIA, OF THE LYCEUM OF NATURAL HISTORY OF NEW YORK, AND OF THE GERMAN ORNITHOLOGISTS' SOCIETY; ETC.

VOL. I. 1859.

" Ibimus indomiti venerantes Ibida sacram,
Ibimus incolumes qua prior Ibis adest."

LONDON:

N. TRÜBNER AND CO., PATERNOSTER ROW.

| *Paris.* | *Leipzig.* | *New York.* |
|---|---|---|
| Fr. Klincksieck, 11, Rue de Lille. | F. A. Brockhaus. | John Wiley, 56, Walker Street. |

1859.

FIGURE 7 The first issue of the *Ibis* (1859), the periodical of the British Ornithological Union.

Tunis markets with Manchester cottons, Sheffield knives, London pickles, sauces, and tinned meat.[147]

Adams's experiences in North Africa also extended to Egypt to study the spread of cholera. He documented many of the birds he encountered along the Nile and published a list of birds of Egypt in the *Ibis*, the periodical of the British Ornithological Union (BOU).[148] Following French engagements in the area, Egypt emerged as an important site in Britain's imperial expansion prior to formal British occupation in 1882. Those vested in empire choreographed "empty landscapes" to colonize for a British audience back home. For instance, the *Ibis* figured the ancient Egyptian bird among the pyramids and the Nile in a land without people (figure 7).[149]

Many BOU members were at the forefront in making Egypt familiar to a British audience. Captain George Ernest Shelley (1840–1910) described the robin redbreast as "confined to Lower Egypt, where it is only a winter visitant. It is as tame and familiar in the sunny climate of Egypt as it is in England, and appears to welcome the stranger, as he sits in the shade of the sont tree, by hopping from bough to bough, and peering inquisitively at him, as though it expected to recognize a friend in the traveller."[150] Shelley, a nephew of the English poet, immediately felt comforted by the presence of the robin redbreast in the North African landscape during his travels in the African continent.

Adams's collection of Egyptian birds revealed time spent in what were becoming the most popular European tourist spots in Egypt, such as Cairo, Thebes, Nubia, and the First and Second Cataracts. When in Nubia, Adams used birdsong as an auditory cue to sonically map a new zoological region. In southern Egypt, where "familiar denizens of the north country disappeared," he encountered "sounds reminding me of Indian jungles," which he identified as the bulbul and the bush thrush.[151] These avian "jungle" landscapes signified "a new ornithic province, the northern outpost of which is Nubia."[152]

While his contribution to the ornithology of Egypt was relatively minimal, his birds were accepted by the British Museum as material evidence of British imperial presence in the region, and as scientific avian trophies in competition with those of France.[153] Alfred Newton, a BOU member, could not hold back his excitement when the Reverend Henry Tristram managed to collect birds in the French colony of Algeria.[154] In a letter to Tristram, Newton exclaimed: "To have carried off such a booty under the noses of French naturalists is a much greater triumph, and the Algerians seem to have expiated all their past cruelties to Christian slaves by the way they have assisted you."[155] Despite Britain's dependence on France for geopolitical control in the Mediterranean, British naturalists attempted to assert their own national and scientific superiority in the region with the accumulation of avian specimens.

## New World Wilderness

Adams solidified his notions of difference between tropical and temperate zones when stationed in the colony of New Brunswick, British North America. When Adams first learned of the new posting to British North America, a comrade asked him: "Where is New Brunswick?"[156] Adams confessed that he had "a rather vague notion of its whereabouts," and therefore fetched "the fine old Imperial Atlas" and traced "out the limits of New Brunswick."[157] In

1866, Adams sailed with his regiment on board HMS *Simoon* and landed in Saint John "to assist in repelling" the Fenian invasions threatening British North America at the time.[158] Adams reached the capital in mid-April and experienced a frigid arrival in comparison to that on the semitropical "shores of the Mediterranean."[159] Adams observed that "the snow had scarcely disappeared, and the noble river, flooded by up-country thaws, was pouring its gelid waters into the Bay of Fundy."[160] From Saint John, the regiment moved on to Fredericton, where it was housed at the Fredericton Exhibition Building. Adams must have lived at the "Pavilion" on lower Regent Street, which housed the married officers such as Adams, who had traveled from Malta with his wife and son.[161]

New Brunswick was relatively unfamiliar to the British imperial imagination in comparison to its "sister" maritime colony of Nova Scotia, where many British military officers such as Captain Thomas Wright Blakiston, discussed in chapter 2, contributed to the knowledge of birdlife around Halifax.[162] New Brunswick emerged as an important site for British investment in the timber industry, and the timber developers included the well known British ornithologist Henry Dresser, who managed the family mill business and shipyard at Lancaster Mills in Musquash during the 1850s and early 1860s.[163] British colonial officials represented the region as a wilderness in the "New World." Lieutenant Governor Sir Arthur Gordon published a chapter titled "Wilderness Journeys in New Brunswick," which described the colony as "one of the least known dependencies of the British Crown," and included "a few sketches of forest life" and descriptions of "natural objects."[164]

Adams's military life in New Brunswick exemplified the typical British North American masculine wilderness experience of sportsman hunting, angling, and canoeing, underscored by encounters with First Nations peoples.[165] New Brunswick provided a site for Adams to reflect on the impact of development, trade, and deforestation on the gradual extinction of certain animal species. He attributed the destruction of "wilderness" to "human agency," but more specifically to the colonizers and their "wanton love of destruction, in many instances similar to that of the Indian, as if a spice of the old savage nature still lurked in them also."[166] The use of the "savage" trope reflected his own essentialist ideology on the degeneration of the white British race and its associations with maintaining "mental culture" for the preservation of the Anglo-Saxon race.[167] However, Adams also encouraged study of "the natural history of its aborigines," treating the Mi'kmaq and Maliseet peoples as natural history "specimens" and tracing their "race characteristics."[168] He concluded, "The Indians of New Brunswick furnish a good illustration of a people rapidly progressing towards extinction, without having preserved any written or monumental record."[169]

Birds such as the "Great Northern or Red-throated divers" emerged as "characteristic objects on almost every New Brunswick lake during the summer months."[170] One species that seemed out of place in the "ungenial weather" was the "Ruby-Throated Hummingbird," which made its home in the "wild woods" of New Brunswick during the summer.[171] Adams noted the "disappearance of the bird commonly known as the Labrador duck (*C. Labradorius*), from the Bay of Fundy and other portions of the adjoining coast," which could not be explained.[172]

While Adams gained knowledge of the local avifauna from colonists in New Brunswick, his ornithological expertise also developed as an accumulation of environmental knowledge from former colonial stations. Considering himself a seasoned traveler, "who has sojourned on the Continent of Europe,"[173] Adams viewed the New Brunswick landscape as flat, declaring that it conjured memories of his disappointment "on the Nile, when, in the absence of monuments of antiquity, he . . . [was] continually surrounded by mud banks, and patches of cultivation, or the eternal sameness of the desert."[174]

After three years' residence in North America, Adams could describe the various aspects of birdlife in New Brunswick and "epitomize a few facts" on the distribution and migrations of birds from "the Old and New Worlds" based on his previous encounters in Scotland, northwest India, Malta, and Egypt.[175] He perfected his naturalist techniques by establishing a "Naturalists' Calendar," which in part showed "the changes of Climate,"[176] the arrivals and departures of each species of migratory birds to New Brunswick. Over three years, he "noted regularly the chief meteorological changes, and also the arrivals and departures of the migratory animals."[177] Adams's transimperial travels allowed him to comment on ideas of bird migration that he had formulated in Malta. He observed that on the European continent, "the migratory birds lag longer on their way north in spring than they do in autumn, whereas in Canada the very reverse would seem to prevail."[178] In 1869, Adams and the Twenty-Second Regiment left New Brunswick and became the last of the many imperial regiments that had been stationed in the colony.[179]

## Temperate Martial Masculinity and Semitropicality

In the works and travels of surgeon-naturalist Andrew Leith Adams, ideas of a temperate martial masculinity reveal distinct origins in the Scottish medical tradition and its circulation to India and the Mediterranean. Adams's early years in Aberdeenshire (Banchory and Aberdeen) taught him the importance of collecting specimens in the "field," which provided lasting impressions of his childhood home in northeast Scotland, especially when he was stationed in India. His formal training and naturalist traditions centered on the Scottish medical community and the Army Medical Department. Unlike Blakiston, who trained as an officer with the Ordnance Department, Adams's notions of a temperate martial masculinity began with the Scottish temperance movement in northeast Scotland and his liberal medical education at Marischal College at the University of Aberdeen.

As a military surgeon, Adams concerned himself with the maintenance of the military body and military fitness in different climatic regions of the British Empire. His transient military career and encounters with differing avifaunas allowed him to trace the contours of tropicality and the temperate, which emerged "transimperially" as Adams moved from one imperial site to the next.[180] When he was stationed in India, his fieldwork as a form of bodily management took on new significance in his attempts to maintain British military and racial health, demonstrating how "place" (i.e., tropical India) incited anxieties over the degeneration of the white, British "race." The resulting collection of birds from the Himalayas reflected imperial territorial interests defined by the Queen's Army rather than the East India Company.

After Adams had returned to Europe, his experiences in Malta allowed him to conceptualize the British Mediterranean station as a moral "semitropical" site for the military officer en route to and from Egypt and India. Here, he studied the effect of winds on the body and documented the migration of hundreds of species during their seasonal migrations from Europe to Africa. For Adams, semitropicality engendered a transitional zone between the temperate and tropical climates of Asia and Africa and represented a landscape in the summer as "bare, weather-beaten, rocky . . . and scarcely a tree to be seen anywhere," and full of verdure and birdlife in the winter. Here, too, he found intermediary peoples: southern Europeans and North Africans whom he viewed as semicivilized.[181] This was also a place where Adams's zoological, as well as his geologic, investigations helped to sustain territorial interests in the Mediterranean region and extend the boundaries of informal empire into North Africa.

Adams's climatic ideologies coalesced when he traveled to New Brunswick and experienced the British North American wilderness and northern climate in relation to his other experiences in Britain, India, and the Mediterranean.

There, he could make comparisons between "Old" and "New Worlds" and assertions about the bird distributions from the four continents where he served and traveled. Adams concluded, "He who has to fight against the climate of Canada on the one hand, and Central Africa or India on the other should be fully developed" before the age of twenty-five.[182] Adams's long career in the Army Medical Department and his contributions to natural history eventually led him to a professorship of natural history at Trinity College, Dublin, and later Queen's College, Cork, after his retirement from army life in 1873.[183] When Adams died in 1882 from pulmonary tuberculosis, he was memorialized as "a self-made man; his advancement and position were essentially his own making."[184]

CHAPTER FOUR

# Leonard Howard Lloyd Irby
*British Military Ornithology on the "Rock"*

Avian Vignette: The Golden Oriole

In May 1870, Lieutenant Colonel Leonard Howard Lloyd Irby of the Seventy-Fourth Regiment noticed bright golden orioles (*Oriolus oriolus*) in his garrison garden at Gibraltar. The orioles in Irby's garden stayed with him all of May. He noted that when his Japanese loquats (*Eriolotrya japonica*) were ripe, the "Golden Orioles remained about as long as the loquats lasted, but would not admit of much observation, as they were very shy and difficult to watch."[1] Irby also mentioned that golden orioles were often heard and not seen. He spent countless hours trying to get a shot at the birds "as they skulked in the thickest foliage of tall trees, continually piping their flute-like note."[2] Many were observed near the mill and the "Second Venta" in the corkwood, and a pair was viewed at the lower part of the first pinewood around the colonial garrison.[3]

Golden orioles used the Mediterranean region as a resting place on their long journeys to and from their winter homes of central and southern Africa.[4] According to the French consul-collector M. F. Favier, the golden orioles crossed the Strait of Gibraltar in great numbers during April and May, returning in July, August, and September in the 1840s to 1860s. Irby agreed with Favier's observation and added that it corresponded to his records on the Spanish side: the first sightings in 1869 on 21 April; in 1870 on 18 April; in 1871 on 4 April; and in 1872 on 11 April. Both Favier and Irby noted that orioles visited fruit-producing regions of North Africa, "where they get the credit of doing much damage to cherries, mulberries, &c.," and therefore were viewed as pests.[5]

As summer migrant birds, flocks of golden orioles were found at Malta, where they sought a "temporary resting-place on Maltese soil."[6] There, according to Andrew Leith Adams, they were often shot, annihilated, and "expelled beyond the precincts of the islands."[7] George Cavendish Taylor of the Ninety-Fifth (Derbyshire) Regiment of Foot viewed a golden oriole in the vicinity of Constantinople when he was sent to the Crimean War.[8] Many naturalists did not know the oriole's wintering grounds, only "the season of

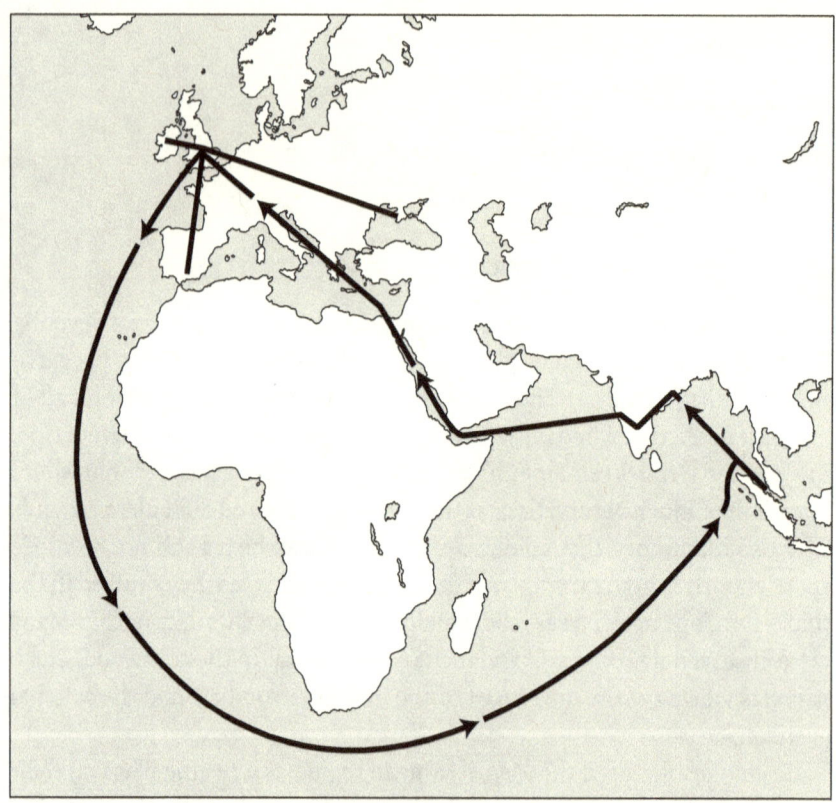

MAP 3  Routes of Leonard Howard Lloyd Irby

incubation in the islands of the Mediterranean, where they assemble in their passage from Northern Africa, but how far they extend on that continent, or how far they pass the Asiatic line, we do not know."[9]

In Britain, the golden oriole, with its stunning yellow and black plumage (male) or somber greenish feathers (female), was considered a rare occurrence, except for a few pairs occasionally breeding in southern England since about 1840 (figure 8).[10] The secretive bird was mostly a summer visitor or an occasional straggler, obtained now and then, but always between spring and fall. Many specimens were taken on the southern coast in April. For example, in April 1824, a young male was obtained at Aldershot, Hampshire, which was purchased and preserved for the Reverend Dr. G. Thackeray, the provost of King's College, Cambridge, and then sent to Eton College.[11]

Most encounters with the golden oriole occurred at zoological gardens and natural history museums across the British Isles. The Zoological Gardens in London received birds from Florence, Trebizond, and Erzeroom.[12] The

FIGURE 8  The Eurasian golden oriole (*Oriolus oriolus*).

bird mounts on display at the South Kensington Natural History Museum (Case 82) included a nest, which was described as "cradle-like . . . skillfully suspended in the fork of a branch," with white eggs, blotched with a reddish-purple tinge.[13] The species was an annual migrant bird to the south of England from "countries south of the Mediterranean in the month of April" and returning in September, but was regularly shot by many taxidermists and collectors to the "deep regret of every right-thinking ornithologist."[14] "This beautiful species," Lord Lilford wrote at the end of the nineteenth century, "is one of the many summer visitors to the continent of Europe, which, as I am firmly persuaded, only requires protection and encouragement to become tolerably common with us."[15]

Currently, the Royal Society for the Protection of Birds lists the golden oriole as a red-status bird, meaning that the species is globally threatened and that its population in the United Kingdom has declined over the last hundred years.[16] However, considering the golden oriole is widespread across Europe, the bird is not classified as an endangered species. Recent efforts by the Conference of the Parties of the Convention on Migratory Species have included the golden oriole as one of the birds set to benefit from the protection of the

Africa-Eurasian Flyway, which would protect the species' wintering grounds in the United Kingdom.[17]

WHEN LIEUTENANT COLONEL IRBY (1836–1905), Seventy-Fourth Regiment (Highlanders), published *The Ornithology of the Straits of Gibraltar* in 1875 and revised it in 1895, he intended the work to assist "officers, who, like the writer, may find themselves quartered at Gibraltar."[18] "For it admits of little doubt," Irby wrote, "that the study of Natural History will always help to pass away with pleasure many hours that would otherwise be weary and tedious during the time military men may have to 'put in' at dear, scorching old 'Gib.'"[19] Irby, a military hero of the Crimean War and the Indian Mutiny, gained status as an intrepid ornithologist who was "sufficiently undisturbed by war's alarms to follow his pursuits over the steppes of the Tauric Chersonese, and again, when called not long after to India."[20] While stationed at Gibraltar, Irby helped to establish the straits as an important site for studying migratory birds and established a network of military men interested in field ornithology.[21]

By the mid-nineteenth century, hundreds of British military officers, such as Irby, had passed through Gibraltar, which emerged as a transimperial site where flows of military bodies, commodities, images, experiences, and ideas circulated to and from other sites in the British Empire. Once known for its naval importance in the Mediterranean with the Great Sieges and the Napoleonic Wars, Gibraltar regained its imperial status with Britain's acquisition of the Suez Canal in 1869, which increased the mobility of its army across the empire.[22] As British military officers made significant contributions to ornithology and the imagining of the Mediterranean, how did their practices and representations of wild birds shape ideas of empire, gender, class, and race in the Strait of Gibraltar? What impact did these configurations have on maintaining the legitimacy of Britain's strategic possession and its monument to empire in the Mediterranean region?

This chapter examines the military and ornithological works of Lieutenant Colonel Irby to understand the ways in which colonial ornithology facilitated territorial maintenance and British imperial place-making in the Mediterranean. In doing so, it focuses again on the "body," as in chapter 3, but extends this analysis to investigate the ways in which the body through performance shaped fieldwork practices and the production of truthful, accurate knowledge of birds in Gibraltar. Geographers have long studied the ways in which images, myths, and symbols are key to imperial place-making.[23] According to Anssi Paasi, special attention should be paid to the geopolitical practices and discourses through which "the narratives, symbols and institutions of national identity are created and how they became 'sediments' of every day [sic] life, the ultimate basis on which collective forms of identity and territoriality are reproduced."[24]

While representational strategies helped shape an imperial imaginative geography of Gibraltar, colonial authority at the Mediterranean station involved the presence of British military bodies as instruments of imperial power and producers of scientific knowledge. Because the British army belonged to the monarch and the government, the British soldier represented a bodily extension of Britain and an essential link to the maintenance of the British Empire. Michel Foucault described the soldier's body as "a fragment of mobile space" that was trained and disciplined to react in a larger sequence of military operations and tactics or performances, which "belonged for the most part to a bodily rhetoric of honour" and exemplified the "deadly military machine" through martial display and spectacle in military colonies.[25] The military body, therefore, involved territorial presence as part of the "basic element of English state spectacle," which altered "moods, social relations, bodily dispositions and states of mind" within the army and in the colonies.[26] According to Ó Cadhla, the scientific performances of Royal Artillery and Royal Engineers officers in Ireland, for instance, reminded local Irish farmers of British atrocities against their people in 1798 and therefore retraumatized them on their lands in the 1840s.[27]

Gibraltar's connections to India as part of the Mediterranean "artery of empire" increased concerns over the effects of racial degeneration on the military body. As early as the 1800s, perceptions of the Gibraltar garrison involved "drunkenness, insubordination, and brutality," whereby "some regiments, fresh from India, and flush of money, were led to excess by the great number of wine-shops allowed in the place."[28] The opening up of the Suez Canal and the reliance on the steam engine and later coal helped to bridge the distance between Britain and India and bring people, commodities, and experiences back home to Britain, a popular theme in the 1840s to 1850s. A contemporary English writer, David Lester Richardson, stated that steam power helped to annihilate "'time and space'" so that "gigantic India and her proud Ruler, small-sized but mighty hearted England, are brought into closer contact and made to afford a noble exemplification of the power of science in the nineteenth century."[29] Here, "The East and West will meet—the swarthy Oriental and the white-faced European will embrace as brethren."[30] Gibraltar's geopolitical positioning in the Mediterranean sustained Britain's hold on its eastern empire and its moral duty to protect it.

Lieutenant Colonel Irby expressed his discontent with European men who were "blackened" from their residence in India. Reflecting on his military service in the Indian Mutiny, he wrote, "Owing to the strong habits of deceitfulness of the natives, no reliance can be placed upon them, if sent out to get eggs. They invariably try to deceive; but their European brethren in trade are often nearly as bad; so that the Asiatic must not come in for all of the black paint."[31] Irby's commentary on the untrustworthiness of Bengali assistants and the negative effects of Asian influences on European bodies raises important matters connecting claims of authority and empirical knowledge. In order to provide trustworthy information, one had to maintain a healthy body and clear mind through physical activity, which offset the perceived damaging effects of colonial service in the British military. The imperial male military body was thus a "site where social structures are experienced, transmuted and projected back on to society."[32] In order to provide truthful information, officers needed to show moral restraint, which tamed "the urge to savagery in themselves" often "associated with the 'primitive' and the 'exotic.'"[33]

By focusing closely on the shaping of military embodiments and masculinity through particular practices *with* a specific landscape (the Rock of Gibraltar), this chapter examines how Irby's approach to ornithology attempted to legitimize Gibraltar as an imperial, noble, and masculine pillar of empire through the collection of wild birds of prey and fieldwork in perilous field sites on the "Rock."[34] However, while Irby embodied a muscular approach to these pursuits, he also demonstrated a masculinity of rational restraint, respectability, and moral considerations surrounding the military body, which eventually extended to moral concerns over the destruction of birdlife in Gibraltar. As Graham Dawson has stated, the fusion of moral and muscular masculinities fostered "a potent combination of Anglo-Saxon authority, superiority and martial prowess, with Protestant religious zeal and moral righteousness."[35]

## The "Rock": Monument to Empire

Gibraltar occupied a sentimental and strategic position in the Mediterranean Sea, overlooking Spain and Africa and securing important trading routes to India. Descriptions, songs, and visual representations portrayed Gibraltar as a masculine protector of the British Empire. The Rock, as a significant landscape, was in itself a monument to the British Empire that was rigorously maintained to sustain ownership and power in the Mediterranean region.[36] The long tradition of British military occupation in Gibraltar helped to naturalize Britain's presence in the Mediterranean region. Gibraltar garrisoned numerous troops from the empire, including "seven thousand men, engineers, artillery, and infantry" from England, Scotland, Ireland, and Canada.[37] One critic exclaimed in Hogg's Weekly Instructor, "The very name of Gibraltar revives in the bosom of every Briton the spark of military ardour."[38] Geopolitical tensions with Spain required a long-term military presence in Gibraltar, which Britain acquired by "British valour" and "preserved" by "statesmanship" with the Treaty of Utrecht in 1713.[39] Together, the Great Siege of 1783 and Nelson's use of Gibraltar during the Napoleonic Wars furthered Britain's image of itself as a superior maritime and military nation.

Many officers viewed Gibraltar as "one of the pleasantest and most interesting quarters a man can have the luck to sojourn in, and service in such a spot" in comparison to the "tropical suns of India or China, the sickly swamps of Demerara, or the wild solitudes of Southern Africa."[40] Despite the cooler climes of the military station, the Gibraltar summers were particularly strenuous for soldiers' bodies, when true masculinity was tested. Summers were "sweltering" and "not so much by the sun's rays as by their reverberation from the bare rock, which becomes almost scorching, and radiates an oven-like heat which is quite stifling."[41] Mosquitoes also abounded and were viewed as "the plague of one's life."[42] Soldiers experienced the levanter, the oppressive easterly wind that "precipitate[d] a clammy and unpleasant moisture" and "paralyze[d] both mind and body."[43] The levanter impressed Irby's friend Royal Engineers Captain Philip Savile Grey Reid (see chapter 5), who illustrated it in one of his sketchbooks. Called by some the "Black Levanter," it affected "man" but also the animals that "move[d] about uneasily," including the birds that "cease[d] their song."[44] "The westerly breezes," which blew "pure and fresh from the Atlantic," were "cool and exhilarating, and both body and mind are invigorated."[45]

British Gibraltar contained numerous military artifacts, such as canons, barracks, and martial street names, which showcased its masculine features for visitors while at the same time reinforcing martial discipline among the troops. Four Russian guns from the Crimean War, presented to Gibraltar by the British government in 1858, still overlook the Mediterranean Sea. British military officers produced and purchased countless visual representations (including both sketches and photographs) that filled military scrapbooks, sketchbooks, and journals to emphasize the British military presence in maintaining the Rock. According to Kathleen Stewart Howe, one of the benefits of military photography involved disseminating the proper positions of military bodies in drills and parades.[46]

Animals entered into the imperial imagination of Britain's claim to the Iberian region. Monkeys, or Barbary apes, were often "associated with the Rock of Gibraltar."[47] Edward Napier (1808–70) of the Forty-Sixth Regiment of Foot described the "standing orders of the garrison" to protect the apes, even though the animals destroyed the fruit and vegetable gardens of the Genoese "who cultivate the western acclivity of the rock."[48] Avian imaginaries also resonated with the Rock, harkening back to the Great Siege of 1783, when the British observed an eagle perched on the westernmost pole of the Signal Station. The sighting was viewed as a favorable omen for the garrison, predicting Britain's victory the following day. This account first appeared in Colonel John Drinkwater's *History of the Siege of Gibraltar* in 1786 and was reiterated in William Henry Bartlett's *Gleanings on the Overland Route* (1851).[49] In 1882, Royal Artillery officer Major Gilbard included the eagle in his booklet on Gibraltar and proclaimed that "the eagle still builds his nest in the crags near the Signal Station."[50]

Of course, not all visitors viewed the Rock the same way. A Canadian soldier of the Hundredth Regiment commented that the Rock of "Gibraltar rises out of the sea like a huge beaver."[51] American opinion tended to be quite critical. For General Ulysses S. Grant, Gibraltar represented the "finest example of *red tapeism* in Europe," as the English occupation of the fortress on Spanish territory adhered to strict "official formalities."[52] One American periodical noted in August 1889, "The whole population of Gibraltar, whether civil or military, is subjected to certain stringent rules. For even a day's sojourn the alien must obtain a pass from the town major, and if he wish to remain longer, a consul or householder must become security for his good behavior."[53] Sentimentalism was used to avert dissension and opposition to the maintenance of Gibraltar on behalf of the British people. Robert Montgomery Martin would employ it in his *History of the British Possessions in the Mediterranean* (1837): "May the day be far distant when treachery or dissension at home shall cause this noble fortress, the protector of our flag, honour and trade in the Mediterranean, to be neglected or contemned."[54]

Gibraltar's classical image as one of the "pillars of Hercules" often excluded non-British locals who were prevented from gaining citizenship and were regularly portrayed as degenerates and "aliens" in Gibraltar's crowded town center and markets.[55] As early as 1804, Samuel Taylor Coleridge (1772–1834) depicted the Spaniards as a "degraded race that dishonour Christianity," and the Moors as wretches who "dishonour human nature."[56] William Makepeace Thackeray (1811–63) described the "Main Street" as the place where "the Jews predominate, the Moors abound."[57] The *Illustrated London News* published

"Sketches of Gibraltar" in 1876, representing the different residents at Gibraltar, such as the Moors at the market, "the Jews and Jewesses," and a "Maltese Milkman."[58] Scottish soldier John Pindar marched through town with his regiment and observed the appearance of its inhabitants, remarking, "The motley group reminded me of the Streets of Calcutta—Jews, Greeks, Turks, Armenians, Arabs, French, Spaniards."[59] His comment that all were "arrayed in all the fantastic dresses of their countries" illustrates how ideas of racial difference depended in part on visual cues of traditional dress.[60] To this group would be added Hindu merchants who arrived in 1870 after the opening of the Suez Canal.[61]

Absent from the ethnic discourses of Gibraltar was any positive discussion of local Gibraltarians. According to David Lambert, the effacement of local indigenous peoples in these narratives highlighted the Rock as "a place through which British troops pass and perform heroic deeds, rather than a place of continuing residence."[62] If mentioned, Gibraltarians were described as a "mongrel race" with no claims to British nationality based on their Spanish ways. Writers often called them "Rock Scorpions" who spoke the "most extraordinary 'pigeon English.'"[63] As M. G. Sanchez has argued, the stereotype of the "undeserving alien colonials" helped to marginalize Gibraltarians in their own territory.[64]

## Muscular Military-Scientific Performances

Military bird collectors exemplified the scientific and masculine hero through tales of their dedication, reasoning, and dangerous escapades as they climbed rocks and trees to shoot birds of prey or collect their eggs (figure 9). Indeed, representations of the military ornithologist evoked heroic imaginaries, as discussed in chapter 2. In the popular science periodical *Nature*, a reviewer proclaimed Irby—who served in the Crimean War with Thomas Wright Blakiston—as reenacting the work of Hercules, the demigod, by bridging the two continents of Europe and Africa through his ornithological work "perched upon the rocky heights of 'Old Gib.'"[65] Irby's achievement was described as "the feat of our modern hero," cultivating a British audience dedicated to empire and science.[66]

Irby, as an officer from the Royal Military College, Sandhurst, would have been trained in mapmaking, gunnery, and scientific practices of classification and documentation, which all helped to sustain the romance of warfare that was integral to British imperial culture and military masculinity through bodily experiences in the field.[67] This type of embodiment involved primarily

FIGURE 9
A. Thornbury, "Bearded Vulture," published in Irby, *The Ornithology of the Straits of Gibraltar* (1895), frontispiece.

a military, gentlemanly body shaped by ideas of Englishness, imperial superiority, and military ardor, in contrast to the lower-class, urban recruits in the army and to colonial peoples.

Irby's connections to the landed gentry in Norfolk County further differentiated his status within the British army. Steeped in the sportsman tradition, Irby collected birds with a gun as an ideal activity that refined the mind and provided the physical exertion to maintain the muscular masculinities of well-trained officers. His approach to ornithology depended on the killing of birds and the physical comparison of specimens for accuracy, similar to Blakiston's practice, described in chapter 2. As Irby wrote, "The only way to avoid . . . errors is never to include any bird in a list except when actually obtained and identified."[68] The bodies of dead birds presented naturalists with material evidence of their scientific discoveries and trophies of the hunt preserved through taxidermy.

The observation of birds was also understood to make better field soldiers in predicting weather patterns during active service. Sir Garnet Wolseley (1833–1913), a hero of the Indian Mutiny, stressed the importance of attendance to both the mind and the body in his *Soldier's Pocket-Book for Field Service* (1871), as "each reacts upon the other."[69] An old "chum" of Lieutenant Colonel Irby's from India, he believed in the "old farmers' predictions of fine or rough weather" through the observations of birds. "When swallows fly high," he wrote, "expect fine weather," while "sea gulls flying inland or collected there in large numbers are fore-runners of bad stormy weather."[70] One might conjecture that Wolseley's experiences with Irby on HMS *Transit* and in the Indian Mutiny influenced his views on observing birds for military campaigning.

Photography and sketching helped to document an officer's masculine pursuit of birds and their eggs on cliffs and mountains and often centered around birds of prey, such as eagles, buzzards, and ospreys. Lieutenant Colonel Willoughby Verner exemplified the ideal officer-photographer in the field when stationed in Gibraltar in the 1870s. "Certainly one of the greatest joys of life to the successful birdsnester," Verner exclaimed, "is to obtain a record of the places he has visited and the haunts of the wild birds he has watched."[71] Verner devoted an entire chapter to sketching and photography in the field, listing the type of camera equipment and the utility of a drawing over a photograph. "My special joy," he wrote, "was to reach some Eagle's nest and endeavour to delineate with pencil and brush 'what the Eagle saw.'"[72] The resulting sketch illustrated his manly achievement of climbing heights to experience the view of one of his favorite birds of prey, while the photograph provided material evidence of the nesting site of a particular species.

Regular routes in Gibraltar allowed officers to observe and collect birds using telescopes and guns around the Rock. Bird collecting occurred at the Neutral Grounds, "close to the Spanish guard-house on the western side," where an officer could find a plethora of golden plover, redhawks, and ringed dotterels.[73] In this way, the field emerged as a space for a "moral locational discourse," helping to regulate the military body in Gibraltar.[74] The sentry spot at the Signal Station was an ideal location for both noncommissioned and commissioned officers to sight the passage of birds in Gibraltar. E. F. Becher, Royal Artillery, noted that the sergeant at the Signal Station observed

FIGURE 10 "Descent to Nest of Bonelli's Eagle," a reenactment by Willoughby Verner in Irby, *The Ornithology of the Straits of Gibraltar* (1895).

a decrease in the number of migrant birds passing over Gibraltar in the year 1882.[75] The higher reaches of the Upper Signal Station allowed Royal Engineers officer Captain Philip Savile Grey Reid to observe a specimen of *Aquila Bonelli*, a type of eagle, which "breed[s] regularly on the eastern side of the rock of Gibraltar."[76] Officers often listed the arrival dates of birds of passage, such as the ring ouzel (*Turdus torquatus*); "the earliest dates in each year being the 8th of April 1868, 20th of March 1870, 9th of April 1871, 12th of March 1872, 28th of March 1874."[77] Such lists served as both a reflection of the accumulation of specimens and a medium to track the annual migration of birds for officers stationed at Gibraltar.

## Moral Ornithology on "Old Gib"

In Gibraltar, army officials designed domestic sites for improvement of the soldier's body, such as the Soldiers' Institute, the Alameda Botanical Gardens, and the garrison churches to help divert military attentions away from pubs and brothels that were understood to damage military fitness.[78] The Reverend John Coventry at the Scottish Presbyterian Church was greatly interested "in the moral and spiritual improvement of the Presbyterian soldiers to whom he officiates as chaplain."[79] John Pindar described how a "Scottish minister," such as Coventry, could make soldiers "feel the hallowing influences of a Scottish Sabbath home," especially in colonial sites, such as in India.[80]

The Gibraltar Garrison Library, established in 1793, provided an exclusive haven for British military officers to read, learn, and exercise the mind on natural history subjects and served as a model for other garrison libraries in the Mediterranean, such as the one at Malta, as discussed in chapter 3. Officers paid an annual membership fee to use the facilities, which excluded local Gibraltarians. As early as 1829, the *United Service Magazine* proclaimed Gibraltar's library to be the "finest institution of the kind out of Great Britain."[81] Countless natural history books filled the shelves, including Gilbert White's *Natural History of Selborne* (1789) and Prideaux John Selby's *Illustrations of British Ornithology* (1821–34).[82] The library was "an invaluable resource, especially in hot weather," and housed thousands of books, "a large and very handsome reading-room, furnished with most inviting sofas, and supplied with all the principal English and foreign newspapers and periodicals."[83]

Commissioned officers attempted to assert their gentlemanly status through their class-based leisure activities in Gibraltar. Irby espoused a moral ethic of restraint toward bird collecting by not including the "exact location" of certain bird species, such as the "White-tailed eagle," for "obvious reasons."[84] His "undistinguished detestation of the race of 'collectors' and wanton destroyers of bird-life" continually pervaded his work despite his own actions in killing birds "to secure the prize."[85] As Irby stated: "The unfortunate part of ornithology, as at present practised, is that it is chiefly confined to the slaughter of birds, whose skins, when compared and examined by table naturalists, are upon the slightest variation in plumage made into new species, without any knowledge of their habits, notes, &c. Much more can be done by observation than by the gun, and when a bird is destroyed all chance of noticing its habits is destroyed likewise."[86]

Irby paid particular attention to the widespread killing of birds for the millinery trade. In the spring of 1874, he noticed how the population of bee-eaters had declined in Gibraltar and the surrounding areas "on account of their bright plumage to put in ladies' hats."[87] He called this practice "a vile fashion," which implicated "no less than seven hundred skins, all shot at Tangier" for a dealer in London.[88]

Contradictory and hypocritical practice abounded. The narration of an authentic adventurous, muscular masculinity in the field was compromised by the common practice of purchasing specimens from local domestic markets at Gibraltar, Seville, and Tangier to enhance collections, especially when it proved difficult to acquire rare species. For example, Willoughby Verner visited the market in an effort to find the *Crysomitris citrinella*, or "Critil Finch" for his collection, although with little success.[89] At the same time, such British military collectors continually made negative remarks about the Spanish "natives" and their class-based pothunting practices. According to Irby, "Spaniards shoot immense numbers" of starlings "at their roosting places" to make "a very cheap and, it may be fairly said, nasty dish in all the ventrollas in the vicinity."[90]

The military officers viewed the scientific approach to shooting and killing birds as more civilized than the pothunting practices of collecting birds for food. Yet, while disparaging the locals, British military-ornithologists still relied on the assistance of Spanish boys in building up their collection of avian specimens. Edward Napier paid a young "muchacho" for his services in the Spanish countryside. The young assistant helped identify a pair of eagles and was "dispatched to secure the spoils" for "the promise of half-a-dollar in the event of finding the bird."[91] Captain Reid mentioned a Gibraltarian boy named José, who brought him "a nest containing nine eggs. . . . I was hardly pleased at such wholesale plunder and directed him to cease his savages among the genus 'perdix.'"[92] By exclaiming his dissatisfaction with José, Reid demonstrated what Mary Louise Pratt has termed "anti-conquest," narrating his innocence in the pillaging by asserting his English superiority in moral egg collecting in the name of science.[93]

## Informal Empire in the Strait of Gibraltar

Military officers extended the boundaries of fieldwork and masculine feats into Spain through collecting in specific foreign field sites. According to G. T. Garratt, Britain's relations with Spain often centered on Britain's attempts to represent Spain as an old imperial power with little political clout in order not to lose Gibraltar.⁹⁴ The "wilds" of Andalucía provided a favorite destination for the officer-sportsman-naturalist, where "the shooting has the charm of a varied bag, and the freedom to wander where you like, as a rule."⁹⁵ For Irby, the best locality for an ornithologist living at Gibraltar was "the country west of an imaginary line drawn due north from Gibraltar as far as the latitude of Seville."⁹⁶ Captain Watkins, who served in Gibraltar after Canada, remarked that the "wild and beautiful scenery of this part of Spain . . . adds in no small degree to the pleasure of the Ornithologist."⁹⁷ Species lists of Spanish birds attempted to erase competing cultures of nature in the Iberian Peninsula. Irby claimed that the "Spanish lists" of local avifauna, especially Baca's "Aves de Españã," were often "meagre and full of errors" and should not be trusted.⁹⁸

Some Spaniards perceived these officers as "all mad" when collecting birds and eggs in the Spanish countryside.⁹⁹ Spaniards nicknamed a particularly enthusiastic collector as "'El loco'—the maniac" for his fanatical "anger in hunting for such trifles as birds-eggs" (figure 10).¹⁰⁰ Such contemporary critiques highlight what could also be understood as the "unnatural" British relation to nature, exemplified by the culture of muscular adventurism. Willoughby Verner confidently confided to his readership, "Whilst all through my life, whenever I have attained the 'decisive point' in a big tree and felt sure of the nest, I have mentally ejaculated with Scud East."¹⁰¹ Based on *Tom Brown's School Days*, Scud East represented the English boyhood hero who delighted in finding a kestrel's nest.¹⁰² As Richard Phillips has written, "The geography of adventure is a cultural *space* in which identities and geographies are constructed" and where imperial masculinities are shaped.¹⁰³

Morocco was another destination for ornithological expeditions among officers based in Gibraltar (figure 11). During the nineteenth century, Gibraltar was the closest and most important European port to Morocco, linking Britain to northern Africa. The British occupation of Gibraltar required continual relations with Morocco in order to guarantee the garrison's food supplies, especially meat.¹⁰⁴ There, British military officers could hire "one or two Moors . . . to pitch tents, load and unload packing animals."¹⁰⁵ It was also a site where the climate was "splendid and healthy, perhaps better than that of Andalucia; and one quits it with the regret that such a fine country should in these days of civilization be, as it were, utterly wasted, a land rich beyond most in soil, minerals, and natural advantages of all sorts, within four days of England, remaining without any real government."¹⁰⁶ In North Africa, Irby could situate his own moral codes of collecting through the Moroccans' relationship with birds.¹⁰⁷ He understood as "superstition" local

FIGURE 11 "Tzelatza Valley, Morocco," from the watercolor sketchbook of Captain Philip Savile Grey Reid, Royal Engineers. Reproduced with permission from David and Andrew Reid, Private Collection.

beliefs that sheltered particular birds, such as swallows and storks, "from molestation by the natives."[108] This approach to birds differed from the lower-class pothunters of Gibraltar and Spain, and although the Moroccans' restraint was attributed to ignorant belief, it aligned more closely with the moral practices of British field ornithologists.

A favorite site to collect birds was "the vicinity of Tangier," a territory lost to Britain in the late seventeenth century.[109] As early as the 1830s, British agent and consul general John Drummond Hay sent specimens back to the Zoological Society of London (ZSL) in England. He described, in a letter to the president of the ZSL, his frustrations in "sending living animals to England from this place."[110] He claimed it "extremely rare that any vessel touch at Tangier on its way to England," and "only 2 or 3 cases have occurred during the four years."[111] Drummond Hay concluded that one could not "trust the masters of merchant ships to take care of the animals on board."[112] His donations included a snake and a "Bonelli's eagle (*Aquilla bonellii*)."[113]

By the 1870s, Tangier continued to be a unique site for ornithologists. British military ornithologists visited French colonial officials interested in ornithology, a practice highlighting the ways in which Anglo-French relations overlapped in the production of scientific knowledge in the Mediterranean region and beyond, as seen in chapter 3. Irby was most interested in obtaining the manuscript on "Moorish birds" of M. Favier, who died suddenly in 1867 after thirty years' residence in Tangier.[114] He recounted how he met the owner of the manuscript and read it in detail with much disappointment. Irby claimed that Favier's manuscript, "upon perusal, amidst a mass of bad grammar, bad spelling, and worse writing, which cost many hours to decipher, did not contain much information."[115]

Published accounts of travel in Morocco, such as Irby's, also helped to identify sites of resistance in informal imperial zones of the Mediterranean, where areas were "forbidden ground to the European" interested in scientific exploration. Irby described fieldwork in Morocco "to be unattended with any danger near the coast, but not east of Tetuan, in the Riff county, or in the mountainous districts."[116] There, "Moors rolled large stones down the only path leading to the summit" and prevented Irby and his colleagues from "ascending."[117] He attributed these actions to "the lawless character of the hill tribes and their Mahometan prejudices."[118] When Captain Reid traveled in the vicinity of Tangier in 1870, he described how British officers had to travel "under the protection of the Sultan" by having a "Moorish soldier" with them, especially when visiting the village of Euzala, where the "dread of robbers was prominent."[119] These incidents illustrate the precariousness of British military presence in Morocco, especially in Tangier, and the way in which ornithology could provide opportunities for military surveillance.

## A Strong, "United," and Moral British Empire

> The British Empire is not like some amorphous jelly fish or invertebrate of low order of vitality that is about to shed its useless limbs. . . . The process that is taking place is the exact opposite of anything of the kind. Adhesion, not fissure, is the law that is in action. Union, not dismemberment, is the law of democratic progress . . . of orderly and organic growth, "until the whole body politic fitly joined together and compacted by that which every joint supplieth according to the working, in due measure, of each several part, maketh increase unto the building up of itself" as one united realm.[120]
>
> —DALTON, "The Colonial Conference 1887"

A focus on Lieutenant Colonel Irby's embodiment in Gibraltar can help to reveal the role of place in the intersection between British military culture and ideas and practices of ornithology in the Mediterranean. As part of the "artery of empire," Gibraltar, like Malta in chapter 3, was a site for the circulation of military bodies, experiences, and ideas from different parts of the British Empire. Many officers, such as Irby, served in India prior to Old Gib, and brought their ideas of racial degeneration and whiteness to the Mediterranean. Gibraltar shared Malta's semitropical environment, as officers battled the sweltering levanter in the summer and enjoyed milder conditions in the winter. Irby, as a gentlemanly officer, pursued field ornithology as a means to ward off the temptations of military life and to display rational restraint in the production of scientific knowledge. His "ethical" approach to bird collecting also reflected his involvement with the humanitarian network in Norwich, England, which attended to the moral rights of animals, or at least those of interest to Britain.

However, unlike Malta, Gibraltar's landscape included a unique geomorphologic landmark, the Rock, which served as an important monument to empire in the Mediterranean region and a sentimental icon back home in Britain. British military officers, such as Irby, here performed ornithological fieldwork, seeking out wild birds of prey on rocky outcrops to increase their collections and to acquire tangible proof of their domination of the region. On the Rock, Irby simultaneously performed muscular fieldwork and humanitarian restraint, reinforcing territorial presence in Gibraltar; his fieldwork forays also bolstered informal empire in Spain and Morocco. This type of military ornithology was produced, consumed, and circulated among fellow officers stationed at the Mediterranean site, in attempts to adhere vital limbs to a strong, "united," and moral British Empire.

CHAPTER FIVE

# Philip Savile Grey Reid

*Red Coats and Wild Birds on the Home Front*

Avian Vignette: Osprey (*Pandion haliaetus*)

Fleet Pond was a favorite haunt in Hampshire County for many British military officers stationed at Aldershot. Soldiers encountered a variety of animals and waterfowl on the county's largest freshwater lake. One frequent visitor was the osprey, a diurnal, fish-eating bird of prey found on all continents of the world (figure 12). Also known as the fish hawk or sea hawk, the osprey feeds primarily on fish, using its hooked talons to grab its prey. Ospreys have yellow eyes and white heads, with a distinctive black eye stripe that makes them different from eagles. European ospreys winter in Africa and use broad fronts to soar over vast distances, including over the Mediterranean Sea.[1]

In 1872, Royal Artillery officer Henry Wemyss Feilden wrote about two ospreys fishing at Fleet Pond, which at the time consisted of "a sheet of water over one hundred acres in extent, in the Government Lands."[2] The warden, who provided the firsthand account of the raptors, sighted the ospreys "circling over the water and constantly plunging for fish," which he "likened to a cannon-shot striking the water." Ospreys were rare occurrences in the British Isles and would have been a great find among field ornithologists. According to the Department of Zoology of the British Museum, the osprey was "a very rare bird in Great Britain," but it bred occasionally in northern Scotland, "where it is carefully protected."[3]

Andrew Leith Adams included the osprey on his list of the birds of Malta.[4] Adams's friend Charles Wright provided a more detailed account of the birds found on the islands of Malta and Gozo. The *arpa*, or "osprey" in Maltese, made its appearance on the islands during the spring and fall, and was often found on the coast, along creeks, and in harbors, such as the Marsamuscetto Harbour and the Great Harbour.[5] The ones observed and collected during Wright's time in Malta were found in October, including a specimen obtained at St. Julian's on 15 October 1861. The bird "had just caught a mullet, of two pounds' weight, and retired to the shore to feast on it."[6] Ospreys were "prominent objects" on the banks of the Nile and on canal banks when Adams traveled to Egypt from Malta.[7]

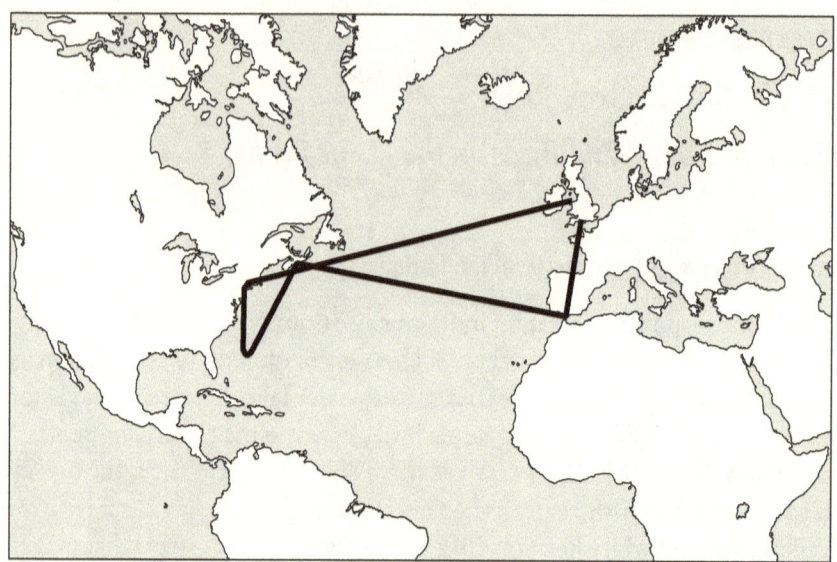

MAP 4  Route of Philip Savile Grey Reid

Lieutenant Colonel Irby described the ospreys at Gibraltar, which he had previously observed in India when serving with the Ninetieth Regiment.[8] According to Irby, when he was stationed in the Mediterranean with the Seventy-Fourth Regiment, the osprey was most abundant in the Mediterranean Straits in winter, yet many pairs nested at Gibraltar in the spring. The Reverend John White was one of the first British naturalists to document the nesting of ospreys at Gibraltar, when he served as military chaplain a hundred years earlier. Irby recorded a pair of breeding ospreys catching fish near Cape Negro, at Lake Esmir, in April, and another pair nesting on rocks west of Tangier. He noted that one pair regularly bred at Gibraltar, "on the rocks a little to the north of 'Monkeys' Cave.'" Irby spent hours in March to July watching the nesting pair, using a telescope from the Europa Advance Battery during the years 1869 through 1871. He described how one brother officer killed an osprey on the wing at Europa Mess House with a pea rifle. "The bird was flying high up over the sea," he wrote, "but the very strong westerly wind blowing at the time caught and landed it among the men's huts; and it now (being well set up) remains a trophy of his skill with the rifle."[9]

Royal Engineers officer Philip Savile Grey Reid, who had served with Irby at Gibraltar, came into contact with "the movements of this cosmopolitan species" when he was stationed in Bermuda.[10] The first osprey he recorded was on 22 April 1874, and another one in 1875, during the same month. They

FIGURE 12
Sketch of an osprey (*Pandion haliaetus*) from Orpen, *A History of British Birds*. Pandion is the name of a Greek hero who was changed into a bird of prey.

were often seen along the south shore of the islands, and at Peniston's Pond. Reid also observed a pair at Devonshire Pond, which led him to the question, "Do they breed here?," based on his experience at Gibraltar when he collected osprey eggs in southern Spain. He believed it was possible considering "little difference of latitude between the two places." However, due to his limited time in Bermuda, he left "the question to be solved by future visitors to the islands."[11]

Ospreys have recently returned to the United Kingdom after being persecuted into extinction in the 1800s.[12] The last recorded breeding of the bird took place in Scotland in 1916.[13] In 1996, the conservation groups English Nature and Scottish Natural Heritage attempted to reintroduce the osprey to central England. By 2007, nineteen osprey chicks had fledged from the

Midlands colony.[14] Their sparse numbers in the United Kingdom are put in contrast to the European population, which is estimated at 8,400 to 12,300 pairs, or 16,700 to 24,600 mature individuals. The osprey is considered a bird of least concern by the IUCN Red List of Threatened Species.[15]

> It has long been the practice of our ornithologists to regard as 'British' any species of which one species has been found in a wild state within the limits of the United Kingdom . . . the fact remains that they ["the American, Asiatic, and European waifs"] are not members of our avifauna, and the young reader should clearly understand that only by a pleasing fiction are they called "British."[16]
>
> —HUDSON, *British Birds*

By the 1880s, British imperial expansion was facilitating a new understanding of Britain's homeland bird life through repeated observations and extensive collections of live and dead birds. With sustained military occupation in the Mediterranean region, and an increased interest in field ornithology, British naturalists had started to trace the connectivity of avian migratory routes from Africa to Europe and, in turn, define Britain's national birds within its own borders. British military officers were instrumental in conceptualizing their "native ornithology," as officers returned to the British Isles from colonial service abroad, bringing with them experiences of different bird species, environmental locales, fieldwork practices, and colonial encounters.[17] Many were members of the British Ornithologists Union (BOU), an exclusive society whose members defined ways of conducting fieldwork and shaped ideas of wild birds during a time of increasing anxiety over industrialism, an expanding empire, the disappearance of the English countryside, and species extinction. As British military officers returned home from service abroad, how did their transient lives and, in particular, their experiences in the British Mediterranean influence their practices and ideas of field ornithology and notions of British birds?

This chapter examines Captain Philip Savile Grey Reid (1845–1915), Royal Engineers, as a homeward-bound officer to Aldershot, Hampshire, to understand how ideas and practices of ornithology circulated back to Britain. Designated as "home of the British Army," Aldershot was an integral site in the transimperial network of military garrisons across the British Empire, connecting England to the Mediterranean, India, British North America, South Africa, and the West Indies. The home station became an important posting for the reunion of family, friendship, military, and ornithological networks in England; its location in Hampshire allowed imperial military officers to ramble in the English countryside and to immerse themselves in what David Matless has termed the "moral geographies of English landscape," fostering temperate cultures of nature through proper conduct in the collecting and documenting of British birds.[18]

Central to my argument in this chapter is an understanding of transimperial processes in the shaping of British military culture and the designation of national birds. Many works have framed British "preservationism" of rural and natural heritage as a reaction to a post-Darwinian world, where population growth and industrial capitalism created a threat to the natural world and therefore spurred an environmental awareness and concern for the protection of traditional rural landscapes and wild birds in Britain.[19] Scholars of environmental history have focused on the wild bird protection movement in Britain within a national context, ranging from the anticruelty campaign to the rational economic benefits of certain avian species.[20] As Brian Bonhomme has stated, wild bird protection in Britain at this time reflected "many things to many people."[21]

Others, such as literary critic Moira Ferguson, have linked national, gendered, and racial identities in the animal protection movement to broader anxieties about the effects of an expanding empire on English domestic society. In her analysis of Sarah Trimmer's book *Fabulous Histories* (1788), Ferguson highlights Trimmer's use of the robin redbreast as a symbol of martial, gendered, and patriotic iconographies of middle-class Anglican conservative culture. "Robin redbreast" defended class interests, exerted authority over rebellions, and resisted rising numbers of foreigners in the capital.

Drawing from recent works in historical geography, I concentrate on the "circuitry of empire" or "web" of networks that shaped "places" and people's experiences across (and beyond) the British Empire.[22] I approach transimperialism in two ways: first as a means to conceptualize Aldershot as a site intricately connected to other military sites in the British Empire; and second, as a way to imagine Captain Reid as a transient figure engaged in multiple localities overseas, especially the British Mediterranean, who influenced field ornithology at home. As Alison Blunt and Robyn Dowling have stated, the spatial imaginary of Britain as homeland is a reflection of "the discourses, everyday practices and material cultures of nation and empire," intimately tied to the politics of place, identity, and collective memory.[23] These imaginaries importantly include avian homeland geographies.

## Aldershot: Imperial Home Station

Aldershot, also known as "the Camp," was established as a garrison by the British government in 1853 to train soldiers and to save money by concentrating troops in heath and moorland rather than in more expensive cultivated areas. The War Department purchased 10,000 acres of land, and its location allowed for easy access to London, Portsmouth, Chatham, and Dover by railway. Aldershot became the first garrison in Britain dedicated solely to the large-scale concentration of troops and emerged as one of the greatest military centers in the British Empire.[24] In 1874, the camp contained up to 754 officers, 15,665 men, and 4,358 horses.[25]

Designated as a home station, Aldershot was an important site in the transimperial network that linked military garrisons across the British Empire.[26] The invention of steam power and the opening of the Suez Canal made possible the effective movement of bringing the "small-sized but mighty hearted England" closer to its colonies, especially India.[27] Homeward-bound soldiers, as English author David L. Richardson stated, would be "twice as efficient as in the olden time, when there were so many obstacles to their breathing a breath of their native air."[28] The Hundredth Regiment spent nine months at Aldershot, "where camp life was on a larger and grander scale," prior to embarking at Portsmouth for the Mediterranean station of "'Old Gib.'"[29] British military officers included images of Aldershot in their travel albums, illustrating its importance in the network of colonial quarters.[30]

As part of the British Empire, Aldershot was subject to the same scrutiny "on the climate" as other colonial garrisons. In a report to the British Association for the Advancement of Science (BAAS) in 1867, Sergeant Arnold stated that Aldershot, "in comparison with any other station, civil or military," cleared a good "'bill of health,'" contradicting popular published statistics that deemed the site unhealthy.[31] Its position within the temperate region of England contrasted with the more tropical stations of the globe and represented "the modest, civilized and cultivated" rather than the more degenerative aspects of hot and humid environments.[32] However, in Britain, popular accounts viewed Aldershot as a "wasteland" or "desert" in the "Maritime Counties," which prompted Sir Robert Michael Laffan, commander of the Royal Engineers of Aldershot (1866–72), to make many improvements in the camp, including the planting of trees and the laying of turf.[33]

Influenced by military reforms in Gibraltar, the education of soldiers at Aldershot was prioritized, and many institutions, such as churches of various denominations, libraries, gymnasiums, and playgrounds, were established to better their lives. The Prince Consort's Library housed books that previously filled the libraries of the hospitals and recreation huts in the Crimea.[34] As mentioned in chapters 2 and 3, women played a key role in the reforming of British military culture at home, as well as in colonial stations. Queen Victoria often visited Aldershot to reinforce her position as the head of the army, as well as her symbolic role as domestic sovereign in nurturing British military men returning home from service abroad.[35] Accounts of her tours of Aldershot were often published in the *Illustrated London News* to demonstrate publicly her ongoing commitment to her army.[36] According to Charles A. Boulton, an Anglo-Canadian officer with the Hundredth Regiment, about 30,000 troops were inspected "by Her Majesty the Queen" as part of a grand review under the Duke of Cambridge in 1859.[37] In Victorian Britain, women were imagined as mothers of the nation, while men were viewed as protectors of Britannia.[38]

Transimperial and Metropolitan (Re)Connections

Born at Welwyn, Hertfordshire, Captain Philip Savile Grey Reid was first posted as an officer of the Royal Engineers at Aldershot in 1865, shortly after completing his training at the Royal Military Academy at Woolwich.[39] Although Reid served at Aldershot three times (1865–69, 1876–78, and 1882–83), it was his return from his second tour of duty that took on new significance. His frequent postings to the home station allowed him to establish and strengthen metropolitan connections with scientific, military, and family networks in England, which helped sustain his gentlemanly English identity, as well as his metropolitan ornithological knowledge and practices. For British military officers, such as Reid, identities were shaped by a transient life across the British Empire, encountering different environments (i.e., climates, flora, fauna), peoples, and cultures. As Alison Blunt has stated, these spatialized, mobile identities were not only "contingent, unstable and decentred" but also "simultaneously grounded, located and contextualized in materially specific ways," exerting particular constellations of power across the British Empire, including at home in Britain.[40]

Aldershot's proximity to London thus helped Reid to reestablish his metropolitan scientific networks in Britain after serving several years abroad and also allowed him to frequent the Zoological Gardens at Regent's Park. On 22 April 1877, he met his longtime friend Lieutenant Colonel Irby "by appointment, in the 'Zoo' . . . to have a farewell yarn with him before he goes out to Gibraltar."[41] They were much amused by the "Snowy herons . . . of the

American ornis," which were "catching blue-bottle flies which settled in their enclosure."[42] As discussed in chapter 4, zoological gardens, such as Regent's Park, served not only as an emblem of empire but also as a place to reconnect with past colonial lives.[43] Prior to their visit, Irby had already donated living avian specimens to the zoo, such as a Bonelli's eagle from Gibraltar and an imperial eagle from southern Spain.[44]

In London, Reid attended meetings of the BOU, to which he was elected in 1877; such "ornithological credentials"[45] he viewed as "about the 'swellest' ornithological 'thing' in England."[46] As an authority on birds, the BOU promoted the scientific "progress of ornithological science in all parts of the globe" and defined fieldwork and nomenclature standards through the union's periodical, the *Ibis*, in relation to the other competing networks of European and American ornithological knowledge (as well as to emerging colonial networks in places such as India).[47] Reid visited the Tenterden Street headquarters at Hanover Square to pick up "a box" for Irby that included an egg of "the Grey-lag Goose from the Laguna de la Jauda" for himself.[48] He also traveled to many British provincial museums to broaden his ornithological expertise. At the Norwich Museum, he met the curator, William Reeve, and viewed the "Raptores [sic], which are truly magnificent."[49] The Norwich Museum housed many avian specimens (skins, eggs) from "all parts of the world," including Irby's avian specimens from the Crimea and Gibraltar.[50]

As a site of convergence for transimperial officers with similar experiences, Aldershot allowed officers to reconnect with others who had served in different parts of the empire. Captain Reid often met with military friends who shared similar interests and experiences in the British Mediterranean. Irby made several trips to Aldershot to see his old Gibraltar friend. When visiting in October 1876, Irby joined a group of other "Gib" comrades on a shooting party, which consisted of Harrison, Denison, and Reid.[51] Reid's friend from Gibraltar, Lieutenant H. R. Kelham of the Seventy-Fourth Regiment, sent him a package "containing a skin and two eggs (one slightly damaged) of the Cinereous Shearwater (*Puffinus kuhlii*) and a skin and one egg (smashed) of the *Stormy Petrel*, taken by him at Fifla," when the regiment was stationed at Malta.[52] Reid also met other military ornithologists, such as Captain Henry W. Feilden, Royal Artillery, who visited Reid at Aldershot to have "a long yarn about birds."[53]

Central to the maintenance of Reid's gentlemanly English identity were his relationships with his family. When on leave, he returned to his childhood county of Hertfordshire, visiting his sister's home at Hatfield, where he stored his collection of birds. On 31 March 1878, when Reid anticipated a chance of "being suddenly ordered away to fight the wily Muscov," he expressed concern "to prepare for the worst as far as concerned my birds' skins and eggs."[54] He "carried all the balance of [his] collection to his sister's house at Hatfield . . . and stowed everything away snug, ready for action!"[55] Reid also visited his sister-in-law, Mrs. Frank Reid, at Southampton to bid her farewell on her trip to Malta in October 1877.[56] On the same ship to Gibraltar was Lord Lilford, president of the BOU, who met Reid the same day.[57] When on leave from Aldershot, Reid married Englishwoman Amy Prime in August 1878, securing his filial ties to Britain despite living a transient life abroad.[58] However, Reid once reflected that he wished to retire from the British army, as he was "very sick of it," and emphasized how much he wanted to "see the world, especially the world of ornithology, while [he was] still able to walk and ride and shoot a bit."[59] Reid reflected on his ambivalent feelings about being at home while imagining new adventures in distant lands.

## Hampshire Rambles and Temperate Cultures of Nature

Aldershot's importance lay in the surrounding countryside for natural history excursions and rambling, which helped maintain Englishness and foster temperate cultures of nature.[60] In southern England, ideal rurality often centered on familiar, domesticated landscapes, removed from urbanization and industrialization, and reflected the role of England as "homeland" in relation to its empire.[61] In this sense, the observing and collecting of British birds was one way to enact a sense of belonging to the English countryside through an appropriate mode of conduct in the country. As naturalist Charles Kingsley (1819–75) stated at the Royal Military Academy at Woolwich in 1871, officers should engage in a "naturalists' field club" rather than a "laboratory" in order to foster "sound inductive habits of mind, as well as more health, manliness, and cheerfulness, amid scenes to remember which will be a joy for ever."[62] More important, "that habit of mind" involved "the habit of seeing; the habit of knowing what we see; the habit of discerning differences and likeness; [and] the habit of classifying accordingly."[63] They are "not merely intellectual, but also moral habits, which will stand men in practical good stead in every affair of life, and in every question, even the most awful, which many come before us as rational and social beings."[64]

As discussed in chapter 2, the Royal Military Academy at Woolwich, in particular, trained young cadets, such as Reid, in the 1860s to conduct fieldwork and perform taxidermy within the Ordnance tradition. There, Reid might have viewed the collection of birds by Thomas W. Blakiston, of the

Royal Artillery, at the Royal Artillery Institution. Pursuing natural history in the "field," and ordering the landscape with scientific nomenclature, disciplined both the body and the mind in British military "body culture."[65] This was a key part of training at home in the English countryside, where temperate and moral martial masculinities could be shaped.

Reid's excursions centered on North Camp, Fleet Pond, Wolmer Forest, Pyestock Wood, and Alice Holt in Hampshire for his sportsman-naturalist outings. When on duty at Aldershot, Reid made use of North Camp on the garrison grounds, a birding area that benefited from the reforestation initiatives by Laffan. Reid often described the site as a good location for common species and their nests, such as "the Long-Tailed Tit, Mistletoe Thrush, and a Tit."[66] Reid commented: "These North Camp Gardens, the nursery for our young trees and shrubs used in planting about the Camps, seem to be a great resort for birds, a veritable oasis in the desert."[67] A fellow officer and caretaker of the gardens, "a man of the 3rd Battalion 60th Rifles," told Reid of a pair of magpies nested in the "thick scotch fir there."[68]

The British army regulated water levels on the War Department Lands, and Fleet Pond was another site for ornithological activities around the garrison grounds. Reid often observed "waders" or waterfowl on his peregrinations, publishing his findings in the the *Zoologist* in 1877. His visit in August of that year overlapped with his duty of draining the pond for the purpose of destroying weeds that attracted wildfowl, including "quite a respectable gathering of Sandpipers & Waders at times."[69] He noted three greenshanks, several curlews and dunlins, a green sandpiper, a tern, and Ray's and pied wagtails. In September, the pond was "nearly empty," but he managed to shoot a male ruff.[70] The management of War Department Lands through tree planting and pond drainage reflected practices of anthropogenic nature in Hampshire County, which had an impact on the avian landscapes in the region.[71]

Aldershot's location, "wholly within Hampshire,"[72] conjured up pastoral ideals from Gilbert White's *Natural History of Selborne* (1789), a best seller in England for more than a century.[73] White spent many years observing and recording the seasonal cycles of the countryside, including the summer and winter "birds of passage round this neighbourhood," and defined traditional British natural history practices of observing birds in the field. His work promoted a spiritual connection with nature through natural theology.[74] Reid often referred to Gilbert White in his journals, and noted the locations visited by White, but refrained from claiming a similar connection to the natural world. In 1877, Reid "went on to Forked Pond Enclosure to look up the

Herons' nests" at Wolmer Forest, and stated, "Old Gilbert White does not allude to it in the 'Natural History of Selborne.'"[75]

As discussed in chapter 4, White's understanding of the local Hampshire avifauna relied importantly on observations and collections of birds from the Mediterranean region, where White's brother, the Reverend John White, served as chaplain to the British army at Gibraltar. In terms of its "migratory connectivity," the "place" of Hampshire can be understood as "'local at all points,' while being definitely, unstoppably translocal."[76] The comings and goings of the county's avifauna also often overlapped with the regular departures and absences associated with numerous naval and military men of Hampshire at Aldershot, Portsmouth, and Southampton.[77]

While in the Hampshire countryside, Reid encountered a "rural system" of English folklore names, and tensions between competing cultures of nature. At Wolmer Forest, a locality in Gilbert White's *Selborne*, Reid showed his agent, "old Goddard," the evasive "Dartford Warbler in the furze, telling him that a nest and eggs thereof was a '*sine quâ non*.'"[78] Goddard "recognized the bird as the 'Black Nettlecreeper' of his rural system, but did not seem to know anything of its nesting habits," and Reid informed him on the species.[79] It was at Wolmer Forest that Reid worried about the fate of the rare Montagu harrier and whether his "friend Roake" would use "his murderous old guns" to kill it. Reid's comments reflected his own justifications for killing the species, as a gentleman of science.[80]

Reid's temperate approach to the killing of birds was legitimized through science and class, but also through race and his imperial experiences in the Mediterranean region. When stationed at Gibraltar in 1871, Reid commented on the arrival of the common quail from England in March and on "the pothunting natives" who "shoot them whenever they see them so that few young ones are hatched."[81] He also recounted his dismay when losing a live Mediterranean gull (*Larus melanocephalus*) that he left at the lodgings of fellow Royal Artillery officer Henry Denison. According to Reid, "The door leading from Denison's garden into the street had been left open by the servants and the bird had doubtless wandered into the road and been picked up (and probably eaten) by some brute of a 'Scorpion.'"[82]

The rhetoric of wasteful destruction and cruelty to avian life was used by the British scientific community to arouse "public opinion," promoting a more temperate nature both at home and abroad.[83] As a member of the BOU, Reid formed part of a group of leading naturalists influencing ideas on the wild bird protection movement and notions of "the British bird."[84] For Reid, homeland birds required protection at home. Though Reid revered homeland birds in his transimperial travels, he became more attached to them after returning home from service abroad. When approaching the English coast (Scilly Isles and Cornwall), he noted, "The Petrels were replaced by *Larus fuscus* in fine plumage," signifying that he had "landed once more on English soil."[85]

## British Homeland Birds

To facilitate the protection of Britain's national birds, practitioners required a definition of what counted as a British bird species.[86] The BOU, through its members, exerted authority concerning the proper documentation of the birds amid different versions of practice circulating in the nineteenth century (other "wrong" versions might designate rare and accidental visitors as British, or ignore scientific nomenclature). As Richard Sharpe, a senior assistant in the Department of Zoology of the British Museum, stated in his *Hand-book to the Birds of Great Britain*, "The story of our native birds [has been] told by a hundred authors in a hundred different ways."[87]

In order to determine the physical boundaries of Britain's birds, a census of the birds of the British Isles was required.[88] In 1861, BOU member Alfred Newton advocated for "a general zoological census" in order to shed light on "an understanding of the geographical distribution of species" and, in particular, "the distribution of British birds" that would have a "design" similar to the "human census of the British Empire."[89] As Newton indicated, "a census of our birds" could only be accomplished "by the co-operation of nearly all the ornithologists in the country."[90] English naturalist Alexander Goodman More (1830–95) devised a census scheme in the *Ibis* in 1865 to categorize the geographic distribution of British birds during their nesting season, "that being the only time when the birds can be treated as stationary."[91] Based on H. C. Watson's notions of botanical districts in *Cybele Britannica*, More grouped different species into "types of distribution" or avian "districts," such as the British, English, Germanic, Atlantic, Scottish, and Highland types.[92] Naturalists stressed the importance of climatic zones in the creation of faunal regions, following the zoogeographic work of Alfred Russel Wallace.[93]

Popular accounts of British birds had been imagined, which involved linking birds to the "temperate zone."[94] In John Gould's *Introduction to the Birds of Great Britain* (1873), Britain's avian landscapes embodied "a temperate climate," with "numerous islets, its rocky promontories and extensive marshes, its natural forests and heathy expanses."[95] According to Gould, "The country a bird resorts to for the propagation of its species should be regarded as its true habitat."⁹⁶ The emergence of avian "climatic races" resulted from the perceived effects of environmental conditions or habitats on living organisms.⁹⁷ As Andrew Leith Adams stated, ornithology provided the medium to determine "which species may undergo transmutation without losing its identity."⁹⁸ Adams concurred with Gould, noting that the habitat "best suited to the bird's constitution is that in which it rears its young."⁹⁹ When Captain Reid observed his homeland birds in Gibraltar, however, he projected his colonial views of the influence of climate onto the species' "habits" or behaviors. When riding home from the Spanish countryside, Reid saw a number of European robins or robin redbreasts in the Andalucían corkwoods, which he described as utterly "unlike their brethren in England . . . their more civilized relations in the north."¹⁰⁰ In Spain, the European robin was crude and slothful.

In 1874, BOU members Alfred Newton and Howard Saunders defined a species as "'British' when a single authenticated example [was] proved to have been obtained in our islands without suspicion of artificial introduction."¹⁰¹ They condemned the "lax method, adopted by older writers on British Ornithology, of admitting any chance straggler from distant lands to a place beside the real inhabitants of this country."¹⁰² The new definition of British birds included "those which belong to that great zoo-geographical region of the Old World of which the British Islands form a portion."¹⁰³ In 1883, the BOU published *List of British Birds* (1883), which included "376 species as the ascertained number of British birds."¹⁰⁴ These species were divided into categories, including residents, summer and winter visitors, and occasional visitors.¹⁰⁵ However, as Hampshire naturalist J. E. Kelsall stated, "To a lover of birds, the difficulty of grouping them as 'resident,' 'regular winter visitor,' &c., is considerable; such fairy creatures prefer to be independent, and seem to rebel against the hard and fast lines of our classification."¹⁰⁶

County lists formed part of the British bird surveys, which prompted a profusion of books dedicated to Britain's avian counties, such as *The Birds of Norfolk* (1870), *Notes on the Birds of Northamptonshire* (1883), and *The Birds of Oxfordshire* (1889). The proliferation of county lists reflected the rise in provincial natural history societies and the ability to transform a rural system into scientific currency through lists and specimens. For such a scheme to be successful, a network of competent observers was required.¹⁰⁷ Captain Reid was known as an ideal collector of "native fauna and flora,"¹⁰⁸ but also for his expertise in noting the differences between Old World and New World species from his field experiences in Bermuda, where he met eastern bluebirds, cardinals, and scarlet tanagers.¹⁰⁹ J. E. Kelsall, in his "list of the birds of Hampshire County," acknowledged the contributions of "Capt. Savile Reid"[110] among other officers, such as Captain Hadfield and Lieutenant Colonel Irby, who helped him to compile a list of birds based on "experience" in the region, a county known to be "rich in birds."[111] Reid's contributions to the birds of Hampshire centered on his collection of birds from Aldershot and Wolmer Forest. Species included the dunlin, a winter visitor that Reid collected at Fleet Pond.[112] His collection also comprised the "Ruff," "Green Sandpiper," "Greenshank," "Curlew," and "Grey Plover."[113]

In Hampshire, Reid searched for the elusive Montagu harrier, a bird named after British military officer and ornithologist Captain George Montagu (1751–1815).[114] The bird itself, a migratory species, was "at one time the most numerous species of Harrier in the fens of the Eastern Counties," but by the nineteenth century its population had declined so greatly that some believed it "shortly to become no longer indigenous."[115] The species often was viewed in Egypt, where it occurred in migration," and was "abundant in the highlands of Abyssinia in winter and spring."[116] In the vicinity of Gibraltar, the Montagu harrier was known vernacularly in Spanish as "Cenizo,"[117] which also applied to hen harriers; but it was often "not met with, except on passage"[118] or in breeding colonies in Morocco, near Lexir. Reid was ecstatic to hear about "several Buzzards (Montagu's Harriers) . . . seen by officers who were shooting at Wolmer," including one around "the north-west side of Brimstone Enclosure" and others "near Lynchboro Pond."[119] In September 1877, Reid hoped that a dead bird from a forest fire was that of the infamous harrier. He stated in his field diary, "As far as I can judge it is undoubtedly a female Montagu Harrier. I brought home the head, sternum, other bones" in order to settle the question "in due time."[120]

Another species of interest from the county was the Dartford warbler, a bird "found on many of the commons and heaths of the southern counties in England," and "one of the most local birds."[121] The warbler, "commonly accredited with a mild climate," was difficult to collect due to its nesting habits in thick furze "on heaths and commons."[122] Both Reid and Irby attempted to find the Dartford warbler during their field excursions in the Hampshire countryside. Reid expressed his disappointment in not obtaining "a nest of the Dartford, in spite of all the trouble"[123] he had "taken in showing the birds themselves and explaining their nesting habits" to his agents. Reid called one of them a "stupid oaf" for forgetting "his promise" to collect the elusive warbler.[124]

Irby eventually found a specimen and donated it to the Natural History Museum for the *Nesting-Series of British Birds* at South Kensington, which aimed to teach the British public about their national birdlife.[125] The new exhibition in 1882 was based on the collection of groups of birds, nests, and eggs in their natural state from England, Scotland, and Ireland, with each of the 159 species displayed in separate cases in the museum's bird galleries, demonstrating their connection to the British landscape. In most instances, the nests were "exhibited with the actual tree, rock, turf and other support which was found with them," as well as the details of the collector, the county, and the date.[126] Irby's Dartford warbler was described as a "local resident in the south of England and more rarely met with in the valley of the Thames and in some Midland counties."[127] It was displayed alongside other species, such as the robin redbreast, one of Britain's "most familiar and characteristic resident species . . . where legendary associations and its fearless nature have combined to make it a general favourite."[128] The robin redbreast also embodied imperial aspirations, as one reviewer in the *Magazine of Natural History* described its song as "self-gratulation," which might, perhaps, be interpreted using William Cowper's words: "I am monarch of all I survey, My right [here] is none to dispute, From the centre all round to the sea, I am lord of the fowl and the brute."[129]

Both Reid and Irby contributed significantly to the British Museum collection of native birds for the new gallery at South Kensington. According to Richard Bowlder Sharpe, the head of the museum's bird collections, if it were not "for the exertions of [Reid and Irby], the Museum might have waited a long time for such rarities as the nests of the Black-throated Diver, the Hen Harrier, and . . . the Wigeon."[130] While at Aldershot, Reid conceived of an expedition in the Scottish Isles as a means of extending his knowledge of the range of British birds. Reid hoped to visit the "extreme north of Scotland, in Caithness" with Irby, where the moor was "a wild swamp or 'floe' much frequented in the breeding season by wild fowl and gulls of sorts."[131] After his retirement, this became a focal point for Reid's "annual expeditions in summer to the north of Scotland to obtain nesting groups of British birds for the series in the British Museum."[132] Reid's work thus helped to disseminate knowledge of British birds for a general public and also advanced the scientific understanding of avian lives in different regions of the British Empire.

Avian Notions of Nation

The production of ornithological knowledge by British military officers, such as Captain Reid, helped reformulate notions of nation and British birds back home in Britain, especially as officers returned home after tours of duty in the British Empire. Many officers, such as Reid, contributed to the development of British ornithology by publishing books or assisting with the arrangement of British birds at museums, thus helping to shape ideas about domestic birds for a general and scientific audience interested in birds and bird protection at the end of the nineteenth century.

Reid's contributions to his native ornithology included the collection of avian specimens from Hampshire County when he was stationed at Aldershot. The imperial home station was an integral site for the maintenance of temperate martial masculinities in the imperial network of garrisons across the British Empire. From Aldershot, transient officers, such as Reid, could strengthen their sense of belonging to the nation by reacquainting themselves with metropolitan scientific, military, and family networks. Hampshire County also provided space for officers to ramble in the English countryside and exercise their temperate cultures of nature, especially after service in more tropical environments.

Reid's ornithological knowledge of Aldershot, however, importantly included experiences in other parts of the empire—especially in Gibraltar—illustrating the connectivity of the British avian landscapes to other places. Officers stationed in the Mediterranean had started to piece together the migratory routes of certain birds, as well as an understanding of the effects of climate on avian species, such as the European robins, in the semitropical environment. Such contingent knowledge underlines the importance of examining the transimperial formulation of ornithological and environmental ideas.

CHAPTER SIX

# Military Ornithology in Place
*Territoriality, Situated Knowledges, and Heterogeneities*

As regiments secured colonies and trade routes overseas, the British military embodied national and imperial power following the successes of the Napoleonic Wars and colonial expansion in Asia, North America, Africa, and the South Pacific. With unique flora and fauna new to science, British military sites provided ample opportunities for naturalist activities in the war zones, colonies, and informal sites of empire. The privileged and well-trained officer mastered not only cartography, gunnery, and fortification but also scientific practices of classification, documentation, and travel writing, which all helped to sustain the romance of warfare integral to British imperial culture. In the imperial imagination, officer-naturalists figured as transient men, circulating with their regiments "in a large orbit . . . in moving from post to post, from province to province, round the belt of the British possessions which encircles the globe."[1]

Historical and cultural geographic research has offered a rich body of literature to unravel the textures and complexities of mobile lives and transimperial movements. This book has explored the intersection between British military culture and ideas and practices of field ornithology in the middle decades of the nineteenth century. It has centered on the British Mediterranean as a site of convergence for transient British military field ornithologists and for migratory birds, tracing the life geographies of British military officers who pursued field ornithology and analyzing the avian imperial archive (from military travel accounts to avian specimens). By paying attention to situated knowledges and place in the production of geographic knowledge, the book has attempted to uncover the ways in which British military ornithology produced the British Mediterranean as a militarized, moral, and zoological region for the benefit of Britain's global empire. The British Mediterranean was not just a single geographic unit or a two-way movement between metropole and periphery, but a series of networks (human and nonhuman) connecting people, birds, and places across (and beyond) the British Empire.

Concentrating on British military encounters with wild birds, this work has traced the spatiality of the production and accumulation of imperial

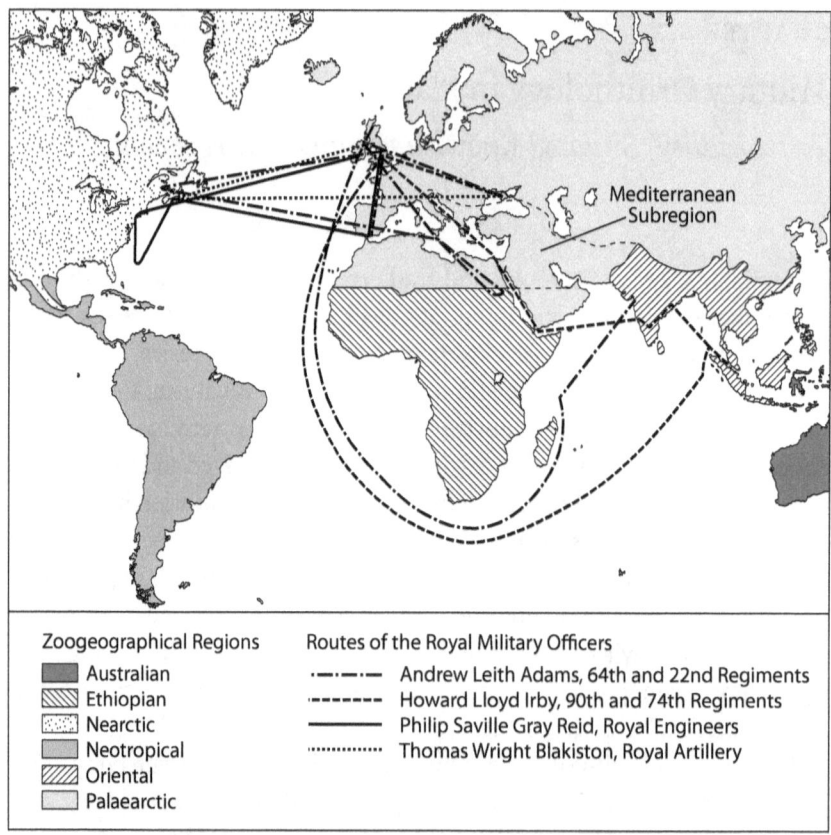

MAP 5  Routes of the royal military officers

environmental knowledge in different geographic locations. Live and dead birds were conceptualized as actants in the formation of officer-naturalists' life geographies, and their material presence illustrates how British military officers negotiated their identities and ornithological knowledge in relation to the various species they encountered and collected. The bodies of birds helped to materialize "diverse intersecting social worlds,"[2] revealing the relational flows of people, birds, and places in the shaping of transimperial military geographies in the Mediterranean region.

Territoriality: Mediterranean Subregion

A major finding of the book was uncovering the contributions of British military officers to the emergence of zoogeography as a branch of biogeography concerned with the distribution of animal species across the globe, and a natural science that was coconstituted in imperial geopolitics. According to Nancy Jacobs, colonial ornithology was considered a "non-instrumental science," which "yielded only small stakes in terms of extraction or governmentality" in the grand scheme of empire.[3] When framed within imperial geopolitics, however, my research revealed how avian zoogeography helped to sustain military fitness during war and times of peace. The protection of the "empire route" to India depended on the efficiency of military manpower stationed at key sites in the Mediterranean, including Gibraltar and Malta. The disciplining of the military body through natural history fieldwork was one way to ensure the territorial presence in the region.

Territoriality was also expressed in the bodies of dead and live birds, as British military officers collected type localities to demonstrate imperial and moral presence and to claim British migratory birds as their national possessions in the Mediterranean. In so doing, the network of British military ornithologists stationed at Gibraltar and Malta helped leading British naturalists to determine the zoological boundaries of the Mediterranean subregion. The emphasis on locality required the production of on-the-spot field knowledge, which was intimately tied to imperial place-making. The new ways of viewing the Mediterranean through the geographic mapping of type localities made visible the avian landscapes that linked zoologically North Africa to Europe. Such mappings also helped to produce an imperial imagination of the Mediterranean as a moral semitropical subregion for the habitability or cultural acclimatization of temperate British officers in transit to and from India.

As this book has illuminated, the concept of "region" *with* "nature" became significant as a tool for contemplating the limits of imperial expansion. The flows of scientific knowledge often gained momentum and power when state intervention defined ways of seeing and documenting "nature" to expand territory and exploit resources. As political geographers have stressed, the ordering of natural landscapes through "spaces of representation," including nature reserves, biodiversity conservation, and, in this book, zoogeographic regions, have been coconstituted by different state and colonial projects, revealing questions of what is "natural" and for "whom."[4]

The management and organization of men and things also extended to British migratory birds as important cultural resources in the maintenance of British military identities. As this book uncovered, the observation of British migratory birds in the Mediterranean region emerged as another way to connect with avian "landscapes of home" when on active duty overseas. Avian bodies provided a site for British military officers to essentialize ideas of Britishness and demonize other cultures of nature, which, for example, saw nothing wrong with eating migratory songbirds in the British Mediterranean. These sentiments were often solidified and gained influence when British military officers returned home to Britain and published works on British birds and assisted in their exhibition at British museums, which, in turn, helped to shape ideas of British national birds. As this book has demonstrated, avian imaginations were entwined with racist, nationalistic, and gendered ideas about particular places and climates.

## Situated Knowledges: Temperate Martial Masculinities

Focusing on the life geographies of four officer-ornithologists revealed the various forms of temperate martial masculinities in British military culture in specific places. The production of the British military scientific hero in the Crimea served as an important national symbol in attempts to gain public support for military expenditure abroad, particularly in the aftermath of the Crimean War, a campaign waged to secure British geopolitical interests in the Mediterranean. As a member of the Ordnance Corps, Captain Thomas Wright Blakiston, Royal Artillery, emerged as an exemplar of the dual role of military life and scientific endeavors in imperial careering. Blakiston's achievements would be memorialized in his published accounts in the *Zoologist* and in his collection of birds of the Crimea at the Royal Artillery Institution, the prestige of which helped to provide him with opportunities to join scientific expeditions to British North America, China, and Japan. The British military-scientific hero demonstrated composure in the theater of war, asserted authority in the field, and amassed scientific trophies of war for a British audience at home.

The British Army Medical Department promoted natural history fieldwork as a necessary pursuit in a transimperial career. Scottish army surgeon Andrew Leith Adams, from the Twenty-Second Regiment, endorsed ornithology as part of a liberal education to help officers maintain a temperate military fitness in colonial stations in different climes. Key to Adams's temperate martial masculinity was the notion of rational restraint and morality, which also extended to a specific climatic zone that shaped an officer's character. The idea of a temperate comportment emerged from his transimperial travels to different geographic regions of the globe and as a by-product of his anxieties over the degeneration of British physical and mental culture in tropical regions. In the British Mediterranean, "semitropicality" involved aspects of both tropical and temperate climatic conditions, the presence of

British middle- and upper-class women, and the seasonal migrations of British birds, all of which helped to regulate temperate martial embodiments and the naturalization of Britain's imperial presence in the region.

Notions of a temperate martial masculinity also were enacted in Gibraltar, where Lieutenant Colonel Howard Lloyd Irby, of the Seventy-Fourth Regiment, embodied both a moral and a muscular masculinity in his scientific fieldwork. Influenced by a humanitarian network in Norwich, England, and his experiences in the "Indian Mutiny," Irby advocated an ethical approach to bird collecting as a way to mediate the "savage" within himself and to ensure the production of objective knowledge of birds. However, Irby also engaged in heroic feats in his pursuit of birds, eggs, and nests on the Rock, reinforcing imperial territorial presence in the Strait of Gibraltar. Here, martial masculinity was explored in connection with a particular landscape in the Mediterranean, revealing how embodiments could be both mobile and place-specific.

By focusing on the military site of Aldershot, the book highlighted how British military field ornithology in Hampshire helped to conceptualize the designation of British national migratory birds as important cultural resources in maintaining temperate cultures of nature, which, in turn, helped to shape ideas of bird protection and conservation in Britain. Captain Philip Savile Grey Reid, Royal Engineers, contributed significantly to the ornithology of Hampshire and ideas of "native birds," providing a moral basis for ideas of bird protection in the British Isles when he was stationed at the imperial home station. Reconnecting to metropolitan scientific, military, and family networks, as well as conducting fieldwork in the Hampshire countryside, helped to sustain Reid's English gentlemanly identity and his temperate martial masculinity.

## Revealing Heterogeneities: Military Life Geographies

A life geography approach in this work was intended not simply to reconstruct the lives of British military officers but rather to investigate the very different ways in which they "negotiated (and [were] negotiated by)"[5] the various colonial settings in which they were stationed and did ornithological fieldwork. Accordingly, the book paid particular attention to their mobile embodiments and the impact of different locales, peoples, and natural habitats on their production of environmental knowledge at each colonial site within the wider British imperial network. The positionality of British military officers in different colonial sites, and their encounters with local indigenous peoples, colonists, and migratory birds, provided an excellent opportunity for interrogating how ideas, identities, practices, and performances of field ornithology emerged transimperially as officers moved from one imperial site to the next.

Certainly, many aspects of their lives overlapped; however, this methodology also permitted a closer examination of the heterogeneity of experiences, networks, and cultural encounters. For example, officers had different regimental cultures and educational backgrounds, and these, in turn, influenced their military identities, fieldwork practices, and networks of trust. Both Blakiston and Reid trained as Ordnance cadets at the Royal Military Academy, Woolwich, but served in different regiments. As a member of the Royal Artillery, Blakiston followed in the footsteps of previous Royal Artillery officer-scientists, including Sir Charles Blagden and Sir Edward Sabine, who were presidents of the Royal Society and maintained a passion for birds. Reid, however, was commissioned to the Royal Engineers, a regiment devoted to its engineering tradition, and which often competed with the Royal Artillery for scientific prestige.

Special attention to biography and place revealed different notions of Britishness and privileged whiteness in British military culture. Adams, for instance, came from humble beginnings in northeast Scotland and sought opportunities as a military surgeon in the British army. His Britishness thus was tied to social mobility within Britain and to his encounters with European East India Company officials in India.[6] As a Scot of modest origins, Adams, however, attained the social status of Blakiston, Irby, and Reid, all of whom came from the English landed gentry. For Adams, life in empire allowed him the freedom to move beyond the social and oppressive conventions of English life, while his home and his homeland birds would forever be tied to the Grampian Mountains of northeast Scotland. He once recalled his "pleasurable remembrances of foreign lands merely by comparison with less agreeable scenes at home, and particularly when contrasted with dismal London fogs and uninviting landscapes."[7]

The notion of the semicivilized southern Europeans and their pothunting practices resulted from multiple cultural encounters in the Mediterranean region and in other parts of the British Empire, including at home in Britain. Blakiston and Irby both engaged with Crimean indigenous Tartars in the Crimean War, as well as French soldiers who practiced pothunting in the theater of war. Many officers, including Blakiston, also sampled nontraditional game birds but retained high opinions of themselves as moral men of science. Adams and Irby formed racialized views of colonial Indians after their own experiences in India during the Second Sikh War and the Indian Mutiny. Irby's involvement in India, in particular, accentuated his prejudices regarding the untrustworthiness of Bengalis. In Gibraltar, British military officers, such as Irby and Reid, perpetuated the stereotype of the pothunting "Rock Scorpion" (a derogatory term for Gibraltarians) as a way to denigrate the locals as undeserving colonials. Similar sentiments were expressed in Malta, where officers such as Adams highlighted the great avian carnage perpetrated by the Maltese. When they were back home in Britain, officers, including Reid, contrasted their temperate cultures of nature with the hunting practices of rural Britons, who engaged in unrestrained destruction of native birdlife. Examining differences and resonances between colonial cultures of nature can shed light on the British military's relationship to nation and empire, as well as ornithological practice.

Following life geographies is revealing, but it is also tremendously difficult to do comprehensively, and this book recognizes its limits. The four military men portrayed here served in different stations across the British Empire, making it nearly impossible to account for all the particularities of place and the role of other networks of empire (e.g., North Atlantic and Indian Ocean) in shaping their identities and their ideas and practices of ornithology. Similarly, while the book attempted to trace the lives of specimens and avian species, it became too challenging to follow the different trajectories and their changing meanings in varying sites and collections. There is potential in georeferencing the avian specimens using Historical Geographical Information Systems to highlight in more detail the movement of specimens and military men in different places and times. Although such work would be time-consuming and labor-intensive, military avian specimens include detailed information on locality, species names, dates, and habitat descriptions, which would allow for the mapping of historical landscape changes, distribution of avian species, and military trajectories of British military men at particular times.

Nevertheless, this book argues that an inductive approach to tracing the life geographies of British military officers using the fragments of the avian imperial archive has provided an exciting, creative, and fruitful avenue of uncovering connections and relations previously overlooked, including the role of the military in zoogeography. Such an approach helped to elaborate and complicate notions of Britain's global empire, the British Mediterranean region, temperate martial masculinities, and British national birds. In these ways, it calls attention to the role of the British military and birds in the geographic tradition and, in turn, the militarization of regions.

# Afterword
## Avian Colonial Afterlives

In October 2009, Michael Palin, the president of the Royal Geographical Society, urged Britons to "stop apologising for their colonial past and be proud of our Empire's achievements."[1] Set within the context of previous apologies for the nineteenth-century Irish potato famine and Britain's involvement in the slave trade by former British prime minister Tony Blair, Palin encouraged his audience to move forward from any perceived crimes of the past as a result of British imperialism. He emphasized the benefits accrued because of empire and concluded: "'We still have links with other countries—culturally, politically and socially—that, perhaps, we shouldn't forget.'"[2]

One unexpected outcome of my research for this book was new insight regarding the impact of colonial cultures of nature on more contemporary issues of bird migration protection in the Mediterranean region, which would not have occurred had I not visited Malta for research in the spring of 2009.[3] Bird hunting is still a hotly contested issue, as evident in the lead-up to the June 2009 European Union elections when I was there in May. The southern European countries of Malta and Italy continue to be portrayed as a misaligned region, where migratory birds are deemed "a harvest, and only one fate would await these birds: the pot."[4] The only refuge for these creatures, according to British writer Michael McCarthy, is in Britain's overseas territory of Gibraltar, reiterating once again the moral righteousness of British cultures of nature in the Mediterranean.[5]

Some have claimed that the EU is another form of imperialism now imposed on the Maltese. Malta joined the EU in 2004, and in 2008 the EU declared a ban on the killing of migratory birds during the spring season. Under the EU Birds Directive, as well as its amending acts, EU countries must seek to protect, manage, and regulate all bird species naturally living in the wild within the European territory of the member states; the directive includes the eggs of these birds, their nests, and their habitats, and the regulation of the exploitation of these species. The member states must also conserve, maintain, or restore the biotopes and habitats of these birds by creating protection zones, maintaining the habitats, restoring destroyed biotopes, and creating biotopes.[6] Although the Maltese government continued to sanction the hunting of the common quail and turtledoves, it reversed its decision after BirdLife International and BirdLife Malta lodged a formal complaint with the EU Commission about the failure to ban the spring hunt completely.

BirdLife Malta has been instrumental in establishing bird sanctuaries, scientific monitoring, and public outreach to educate the Maltese on the importance of bird protection. These conservation efforts have proved fruitful, with a resulting increase in bird populations and the diversity of species. For example, BirdLife Malta has documented the first confirmed breeding bird of prey in fifteen years. Following the 2009 spring hunting ban, a pair of common kestrels successfully bred and raised at least three chicks in the Maltese islands. Furthermore, the host family with whom I was staying for my research described their excitement at seeing hoopoes in Malta, a sight they had not experienced in years. They also described the European robin that visited their mother's bird feeder on a regular basis.

Not surprisingly, the ban on the spring hunt has caused outrage among Maltese bird hunters and trappers who claim that it is their traditional and legal right to pursue spring hunting, which they have done since "time immemorial." Although the Federation of Hunting and Conservation Malta advocates for sustainable hunting practices, there remains great resistance to the restrictions imposed on hunting practices during the spring season. The bird sanctuaries that have been set up to protect migratory birds are often trespassed by bird hunters who kill the protected species as a form of protest. Most alarming are the actions taken by the more resistant group of bird hunters who have uttered death threats, slashed tires, and torched the cars of some of the bird protection advocates. The Federation of Hunting and Conservation Malta even issued a news release titled "BirdLife Malta Infiltrates Education Department," which lambasted the organization for its "indoctrination of children with deceiving material; it is denying us from a just platform on what is the environment and its conservation."[7] For those involved with BirdLife Malta, these types of statements must be extremely frustrating.

As I observed during my short stay on the islands, the Maltese are ambivalent about Britain's involvement in Malta. Malta only gained independence from Britain in 1964 and formally cut ties in 1979 with the removal of British forces by then prime minister of Malta Dom Mintoff. Restrictions on Malta's cultural traditions (e.g., the use of fireworks at local village feasts) are often seen as interference from the EU.[8] While a significant British expatriate community still lives in Malta, I spoke with older Maltese residents about their feelings toward the British: they commented how Malta was never their own during British occupation yet they had good jobs and good pay. Nationalist

FIGURE 13 Ta' Braxia Cemetery, the older section where British military and Royal Navy officers (commissioned and noncommissioned) and their families are buried, Pieta, Malta. Photo by author.

sentiments can be traced back to the 1880s as many viewed Malta as an extension of Italy rather than as a British colony. These views intensified after the British expelled numerous nationalists to Uganda for fear of treason during the Second World War.[9] As Jon Mitchell uncovered in his fieldwork, it is "commonly believed that the British deliberately kept the Maltese poor and ignorant because it served their interests."[10] I could understand the Maltese resentment toward the British colonial experience because of similar sentiments expressed by nationalists/separatists in my own home province of Québec in Canada.[11]

Animosity toward the British was evident during my field research, as many of the nineteenth-century British military heritage sites are now harder to find, taken over by Maltese institutions, or destroyed for housing developments. The cemetery Ta' Braxia, the resting place of the nineteenth-century British colonial elite, now lies overgrown with weeds and filled with broken tombstones (figure 13). Even Queen Elizabeth's former residence, where she lived for several years after her wedding to Prince Philip in 1947, is barely noticeable in a run-down part of town. Ashish Chadha suggests that this "discard, decay and abandonment of colonial memorial monuments in the postcolonial landscape stems from the ambivalent meanings that such a heritage site generates."[12]

Today, hundreds of British birders visit Malta on a yearly basis to observe the migration of birds but also to deter local hunters from killing them. Since the late 1990s, BirdLife Malta has organized a Raptor Camp during the annual autumn raptor migration for both local and international volunteers to curtail illegal hunting activity. As BirdLife Malta states: "This is no bird watching holiday, but a serious conservation effort!"[13] Here, the "modern geopolitical imagination" of the nineteenth-century British Mediterranean has retained some "elements of continuity" in the construction of the Maltese pothunter.[14] As Homi Bhabha has argued, the discursive strategy of colonial discourse is a "form of knowledge and identification that vacillates between what is always 'in place,' already known, and something that must be anxiously repeated."[15] As a result, it ensures the stereotype's "repeatability" and circulation in changing historical contexts.

Derek Gregory has suggested that in order to overcome this type of "colonial amnesia,"[16] we must be made aware of colonial historical geographies. While the subject of birds might seem benign in comparison with the more overt acts of violence of slavery and famine, a self-awareness and sensitivity of one's position in the world as a result of the institutions, practices, and identities that emerged from the British Empire is still needed in order to deconstruct and challenge the "colonial amnesia" of cultures of nature in particular places, such as in Malta. The annual presence of British "moral" birdwatchers as a means to combat the "savage" Maltese pothunter will not resolve the migratory bird hunting issue in Malta—it only repeats a stereotype and enlivens old tensions within a British colonial culture of nature that marginalized lower-class Maltese in the nineteenth century. As I have demonstrated in this book, the stereotype of the colonial Maltese pothunter continues to circulate in Europe. A critical historical geography of empire can help to trace some of the genealogies of colonial cultures of nature in particular places such as the Mediterranean, and to contextualize tensions among different actants in conservation efforts dedicated to migratory birds.

# Notes

## Introduction

1. Zammit, "Malta No Mecca for Migratory Birds."
2. RSPB, "Bringing an End to the Maltese Killing Fields"; Franzen, "Emptying the Skies," 48.
3. Malta also is a signatory of the Convention on Migratory Species, which includes a memorandum of understanding regarding birds of prey of Africa and Eurasia.
4. For an overview of the bird hunting issue in Malta, see Fenech, *Fatal Flight*.
5. Michael McCarthy, "Conservationists Mobilise to Halt Mass Slaughter of Birds in Malta."
6. Michael McCarthy, one of Britain's most popular nature writers, was awarded the medal of the Royal Society for the Protection of Birds. He was formerly the correspondent for the *Times of London* and environment editor of the *Independent*. McCarthy, *Say Goodbye to the Cuckoo*, 20-21.
7. Neumann, "Dukes, Earls, and Ersatz Edens," 79-98; Neumann, "Moral and Discursive Geographies," 813-37.
8. Campbell, *The British Empire*, 146; Garratt, *Gibraltar and the Mediterranean*, 136-37; Farnie, *East and West of Suez*, 115; Angell, *The Defence of the Empire*, 125; Robinson and Gallagher, *Africa and the Victorians*, 13. The concept of the modern world system has been developed by Immanuel Wallerstein in his *Unthinking Social Science*.
9. Burroughs, "An Unreformed Army?," 160-88.
10. Garratt, *Gibraltar and the Mediterranean*, 135.
11. Howell, *Geographies of Regulation*, 153. In her book *Captives: Britain, Empire and the World*, Linda Colley has stated that it is a "region often left out of the history of English and British commercial and imperial endeavour" (17). Other works on the historical geography of the British Mediterranean include Howell, "Sexuality, Sovereignty and Space," 444-64; Lambert, "'As Solid as the Rock?,'" 206-20.
12. McNeill, *The Mountains of the Mediterranean World*; Butzer, "Environmental History in the Mediterranean World," 1773-1800; Davis, *Resurrecting the Granary of Rome*.
13. For example, in 1856 Britain capitalized on the Commercial Treaty with the sultan of Morocco, which opened Moroccan ports to the world market "liberal" trade. See Gambi, "Geography and Imperialism in Italy"; Bourguet et al., *L'invention scientifique de la Méditerranée*; Izzo and Fabre, *La Méditerranée française*; Ben-Artzi, "The Idea of a Mediterranean Region"; Claval, "About Rural Landscapes"; Darwin, "Imperialism and the Victorians"; Morin, "Surveying Britain's Informal Empire"; Fan, *British Naturalists in Qing China*.
14. Curtin 1989, 44; Srhir 2005, 259; Colley, *Captives*, 103.
15. French, *Military Identities*, 3.

16. Thomas, "The Home of Time," 27.

17. Pick, *Faces of Degeneration*, 156.

18. Early groups to preserve rural landscapes included the Commons Preservation Society (est. 1865), the Society for the Protection of Ancient Buildings (1877), and the National Footpaths Preservation Society (1884). See Marsden et al., *Constructing the Countryside*, 76.

19. The Catesby specimens were sketched and published in the 1843 version of William Yarrell's *A History of British Birds*. Mark Catesby traveled to British North America and the West Indies and published his *Natural History of Carolina, Florida and the Bahama Islands* (1731–43). Gibraltar was also where John White, rector of the garrison in the 1770s, suggested ideas about British bird migration from Africa to Europe based on his collections and observations. These ideas, which were published by his brother the Reverend Gilbert White in his best seller *The Natural History of Selborne* (1788–89), highlight the ways in which avian landscapes connected imperial territories and national identities. However, ideas about bird migration circulated even earlier than this account. See Mason, *George Edwards*, 29; Foster, "The Gibraltar Collections," 30–46.

20. Mearns and Mearns, *The Bird Collectors*; Berger, *Science, God and Nature in Victorian Canada*, 11.

21. Allen, *The History of American Ornithology before Audubon*; see also Streseman, *Ornithology from Aristotle to the Present*.

22. In *The Bird Collectors*, Mearns and Mearns identify several British military officers as the largest group of contributors to ornithology in the nineteenth century, as officers observed, documented, and collected birds in Britain's colonies overseas.

23. Recent works on empire and ornithology focus on the American context. See Quintero, "Trading in Birds," 421–45; Kohout, "From the Field." Works on colonial ornithology include Jacobs, "The Intimate Politics of Ornithology in Colonial Africa," 564–603; Jacobs, *Birders of Africa*.

24. Hume, *Ornithologists of the United States Medical Army Corps*.

25. MacLeod, "'Strictly for the Birds,'" 315–52.

26. MacDonald, "What Makes You a Scientist," 58.

27. The scientific study of birds was essential in the conceptualizing of the zoogeographic boundaries of the globe. David N. Livingstone discusses the history of zoogeography but does not mention birds in *Putting Science in Its Place*, 161–62. See also Livingstone, *The Geographical Tradition*; Driver, *Geography Militant*.

28. Jacobs, "The Intimate Politics of Ornithology in Colonial Africa," 565; Allen, *The Naturalist in Britain*, 214.

29. Doughty, *The English Sparrow in the American Landscape*; Doughty, *Feather Fashions and Bird Preservation*.

30. Wilson, "Directing the Flow," 247–66. His Wilson's more recent work includes the book *Seeking Refuge*.

31. Lambert and Lester, *Colonial Lives across the British Empire*.

32. Lambert and Lester, 24.

33. Lambert and Lester, 23–24.

34. For works in historical feminist and critical geographies that attend to the gendered and racialized geographies of British military subject positions in different sites of the British Empire, see Blunt, *Travel, Gender, and Imperialism*; Morin, "Embodying Tropicalities," 23–25; Morin, "Charles P. Daly's Gendered Geography," 897–919.

35. Nash, "Performativity in Practice," 653–64.

36. Raffles, "Local Theory," 324.

37. I use the term "actants" rather than "actors" to move away from human-centered notions of agency. See Whatmore, *Hybrid Geographies*.

38. Gregory Cushman gives birds a central role in his global ecological history of the guano industry in the Pacific world: Cushman, *Guano and the Opening of the Pacific World*.

39. Ingold, "Rethinking the Animate, Re-animating Thought," 14.

40. Massey, *For Space*, 139. Massey's notion of the "event of place" is discussed in her chapter on "the elusiveness of place." She provides examples on the ways in which mobile lives, such as migrating swifts, contribute to this notion of place.

41. Massey, 130.

42. Alan Lester builds on his earlier work to include nonhuman agency and assemblage theory in these imperial networks: Lester, "Introduction," 1–15. He suggests new research directions in Lester, "Commentary," 120–23. See also Greer, "Zoogeography and Imperial Defence," 454–64. Scholars who have drawn from Alan Lester and David Lambert's circuitry of empire include Beattie, Melillo, and O'Gorman, "Eco-cultural Networks in the British Empire"; Jacobs, "Africa, Europe and the Birds between Them," 92–120.

43. See Gambi, "Geography and Imperialism in Italy," 74–91; Bourguet et al., *L'invention scientifique de la Méditerranée*; Izzo and Fabre, *La Méditerranée française*; Ben-Artzi, "The Idea of a Mediterranean Region," 2–15; Claval, "About Rural Landscapes," 7–24.

44. As geographers have demonstrated, each site—colony, port, battlefield, troopship—within the imperial network fostered "'its own possibilities and conditions of knowledge,'" and sites were connected by "communicative circuits of empire, and could thus be mutually affecting." Alan Lester uses Miles Ogborn's phrase "geographies of connection" in his *Imperial Networks*, 5–8; Ogborn, *Spaces of Modernity*, 19. See also Martins, "Mapping Tropical Waters," 148–68; Cannadine, *Ornamentalism*; Lester, "Imperial Circuits and Networks," 124–41; Lambert and Lester, *Colonial Lives across the British Empire*.

45. Febvre, "The Mediterranean Is the Sum of Its Routes," 70; cited in Braudel, *The Mediterranean and the Mediterranean World*, 276. Braudel's work was first published in 1949.

## Chapter One

1. British Museum of Natural History, *Guide to the Gallery of Birds*; Russell, Hansell, and Reilly, "Constructive Behaviour," 77–96. The rationale for inclusion in the nesting series was ambiguous, as although all 159 cases displayed birds recorded in Britain, seven of the species represented are not known to breed in the British Isles. Thank

you to Mark Adams, Douglas Russell, and Rys Jones for information on the nesting series of British birds displayed at the Natural History Museum at South Kensington.

2. See the preface in *Guide to the Gallery of Birds*.

3. "No. 38. Redbreast or Robin (Erithacus rubecula)," in *Guide to the Gallery of Birds*, 14.

4. Royal Artillery Institution, *Minutes of Proceedings of the Royal Artillery Institution*, 56.

5. Anon., "Obituary: Captain George Ernest Shelley, FZS," 93–94.

6. Mearns and Mearns, *The Bird Collectors*, 187–208.

7. Sharpe, "Ornithology at South Kensington," 175. Sharpe joined the British Museum as a senior assistant in the Department of Zoology in 1872, overseeing the department's bird collection.

8. When viewing the actual specimens, one uncovers the various naming and documenting practices used by the different collectors throughout the object's "life." For example, Andrew Leith Adams's collection of birds at the World Liverpool Museum indicates labels with his own personal numbering practices, which he used to organize his private collections in Malta. Another label includes Sir William Jardine's numbering practices, and then Canon Tristram's. What is revealed is how natural history museums themselves are developing their own unique storage organizations according to practicality, as well as taxonomy. While natural history museums use particular taxonomic schemes to organize their respective bird collections (e.g., Moroney, Bock, and Farrand; Peters; Sibley and Monroe), many collections cater to the needs of the museum and its curators.

9. Lambert and Lester, *Colonial Lives across the British Empire*, 23–24.

10. Lambert and Lester, 23–24.

11. For work on British military material culture, see Burant, *Drawing on the Land*; Burant, "Record of Empire," 120–28.

12. Richards, *The Imperial Archive*. Richards includes lists of birds as part of the "imperial archive" but does not mention natural history specimens or photographs.

13. Richards, 15. For the distinction between the "archive" as a metaphorical construction and archives as real institutions, see Schwartz, "Reading Robin Kelsey's *Archive Style*," 201–10.

14. Evans, *Ireland and the Atlantic Heritage* 42, as quoted in Ó Cadhla, *Civilizing Ireland*, 8–9.

15. Evans, *Ireland and the Atlantic Heritage*, 6.

16. Doughty, *The English Sparrow in the American Landscape*; Haraway, *Primate Visions*; Elder, Wolch, and Emel, "Le practique sauvage."

17. Haraway, "Teddy Bear Patriarchy," 21. See also Poliquin, "The Beastly Art of Taxidermy," 11–16.

18. Alberti, "Objects and the Museum," 559–71; Alberti, "Constructing Nature behind Glass," 73–97; Alberti, *Nature and Culture*; Delborough, *Collecting the World*.

19. Henry McGhie's work, in particular, shows the importance of considering the contextual circumstances surrounding the acquisition of objects within a postcolonial museum. McGhie highlights the biography of nineteenth-century ornithologist Henry Dresser to help trace his complex business and scientific networks in his accumulation of ornithological knowledge, and the history of the collections at the Manchester Museum, within the context of both empire and colonialism. McGhie, "Contextual

Research and the Postcolonial Museum"; McGhie, *Bird, Books and Business*. For an interesting interview on the use of avian specimens and biography, see the discussion of the Manchester Museum's bird collection, its origins, and its relationship to the Revealing Histories project in the YouTube video "Jill Malusky and Henry McGhie." See also Greer and Bols, "'She of the Loghouse Nest,'" 45–67; Kohout, "More Than Birds," 83–96.

20. Cameron, "Oral History in the Freud Archives," 38–44; de Silvey, "Observed Decay," 318–38; Mills, "Cultural-Historical Geographies of the Archive," 701–13.

21. Lorimer, "Herding Memories of Humans and Animals," 512. See also Lorimer, "Songs from Before," 57–74; Patchett and Foster, "Repair Work," 98–122; Patchett, "Taxidermy Workshops," 390–404.

22. Forsyth, "The Practice and Poetics of Fieldwork," 128–37; Forsyth, "Biography and the Military Archive," 44–57; Forsyth, "On the Edges of Military Mobilities," 48–50. See also Greer, "Untangling the Avian Imperial Archive," 59–71.

23. Raffles, *In Amazonia*, 127.

24. Livingstone, *Science, Space, and Hermeneutics*. Laura Cameron writes about Stephen Daniels's earlier use of "life geography" in Cameron, "Oral History in the Freud Archives," 38–44.

25. Trevor J. Barnes, "Lives Lived and Lives Told," 409–29.

26. Rich, "Woman and Bird," in *What Is Found There*, 7.

27. Geographer J. B. Harley has been instrumental in using material culture as primary sources in historical geography and viewing maps as "texts" or cultural constructs rather than "a mirror of reality." See Harley, "Historical Geography and the Cartographic Illusion," 80–91.

28. Ophir and Shapin, "The Place of Knowledge," 15.

29. Lorimer and Lund, "Performing Facts," 131.

30. Edwards, Gosden, and Phillips, *Sensible Objects*, 3.

31. Raffles, "The Uses of Butterflies," 514–15.

32. I use Felix Driver's use of Latourian notions on the circulation of geographic knowledge. Driver draws from Latour, *Science in Action*; Driver, *Geography Militant*, 29–30.

33. Driver, *Geography Militant*.

34. Richards, *The Imperial Archive*, 11.

35. Jacobs, "The Intimate Politics of Ornithology in Colonial Africa," 565; Allen, *The Naturalist in Britain*, 214.

36. Browne, "A Science of Empire," 466. For works on biogeography and empire, see Zeller, "Classical Codes," 20–35; Zeller, "Humboldt and the Habitability of Canada's Great Northwest," 382–98; Bewell, "Jefferson's Thermometer," 111–38.

37. Greer, "Geopolitics and the Avian Imperial Archive," 1317–31; Greer and Cameron, "The Use and Abuse of Ecological Constructs," 451–53.

38. Justin Stagl discusses how the accumulation of curiosities led to the specialization of collections (plants, animals) and therefore to becoming more research oriented. Stagl, *A History of Curiosity*, 114. See also Griesemer, "Modeling in the Museum," 11.

39. Sclater, "On the General Geographical Distribution of the Members of the Class Aves," 132. 137.

40. Sclater, 135.

41. Sclater, 135.

42. Wallace, "Letter from Mr. Wallace," 449.

43. Wallace, 449.

44. Browne, *The Secular Ark*; Camerini, "Evolution, Biogeography, and Maps," 700–727.

45. Sclater, "On the General Geographical Distribution of the Members of the Class Aves," 132.

46. Sclater, 132. By the 1870s, Sclater was secretary to the Zoological Society of London; was the founder and editor of the British Ornithological Union periodical the *Ibis*; and was involved with the BAAS and the Royal Geographical Society. He edited the *Ibis* from 1858 to 1913 (apart from the years 1865–77). See Edwards, "Sclater, Philip Lutley."

47. Moreau, "Centenarian *Ibis*," 29.

48. Becher, "Zoological Notes from Gibraltar," 100.

49. Raffles, "The Uses of Butterflies," 524.

50. Browne, "A Science of Empire," 453–75; Laidlaw, *Colonial Connections*, 33.

51. Edney, "British Military Education," 16–17; Edney, *Mapping an Empire*; Braun, "Producing Vertical Territory," 13. See also Peers, "Colonial Knowledge and the Military in India," 157–80.

52. Hannah, *Governmentality and the Mastery of Territory*, 128.

53. Edney, *Mapping an Empire*.

54. Holland and Markides, *The British and the Hellenes*, 6.

55. Johnson, "Type-Specimens of Birds," 181.

56. British Museum, *The History of the Collections*, 398–99. Other museums included the Zoology Museum at Cambridge University, the Manchester Museum, and the World Liverpool Museum.

57. Leonard Howard Lloyd Howard Irby to Philip L. Sclater, 16 October 1869, Library of the Zoological Society of London, London, UK. Some existing letters by Philip P. Sclater are located at a few archives. For the most part, however, the location of Sclater's materials (diaries, notebooks, and personal letters) is still unknown.

58. Livingstone, *Putting Science in Its Place*, 171.

59. Richards, *The Imperial Archive*, 3; Allen, *The Naturalist in Britain* 215.

60. Known as the "Strickland code," the rules drawn up by the BAAS committee became a standard on zoological classification and nomenclature, replacing all other classification systems of the period. The "code" was based on the twelfth edition of *System Naturae*. See Bircham, *A History of Ornithology*, 213–16.

61. Anderson, "Census, Map, Museum," 184.

62. Foucault, *The Order of Things*, 132–38; Ritvo, "Zoological Nomenclature," 334–53; Ritvo, *The Platypus and the Mermaid*, 1–50.

63. Hannah, *Governmentality and the Mastery of Territory*, 56. See also Cohn, *Colonialism and Its Forms of Knowledge*.

64. David N. Livingstone makes the distinction between closet naturalists, who compared and classified specimens in museums, and traveling naturalists, who documented data "on the spot." See Livingstone, *The Geographical Tradition*, 135.

65. Ritvo, *The Platypus and the Mermaid*, 19.

66. For works on "the field" and "field cultures," see Outram, "New Spaces in Natural History," 249–65. See also Hevly, "Heroic Science of Glacier Motion," 66–86; Cameron and Matless, "Benign Ecology," 253–77.

67. Outram, "On Being Perseus," 286.

68. Godlewska, "Humboldt's Visual Thinking," 266.

69. Griesemer, "Modeling in the Museum," 20–21.

70. Griesemer, 20–21.

71. Datson, "Type Specimens and Scientific Memory," 162. Many British military avian eponyms are included in Mearns and Mearns, *Biographies for Birdwatchers*.

72. By the 1830s, naturalists commonly used arsenic soap as a preservative on bird skins to protect specimens from insect infestation and decomposition. Many of these type specimens are still in existence, unlike some of the earlier specimens that were preserved by pickling or drying them. Johnson, "Type-Specimens of Birds," 174. For a history of taxidermy techniques, see Farber, "The Development of Taxidermy," 350–566.

73. Thomas Davies to John Ellis, "On a Method of Preparing Birds for Preservation," paper read on 22 March 1770, L & P/V/165, Library and Archives of the Royal Society; London, UK. Davies, "On a Method of Preparing Birds for Preservation," 34–35. Davies collected birds for his patron, Thirteenth Earl of Derby, in various postings. Some of these birds are now housed at the World Liverpool Museum. See Dickenson, *Drawn from Life*, 189–98; Fisher, *A Passion for Natural History*.

74. Davies, "On a Method of Preparing Birds for Preservation."

75. Raffles, "The Uses of Butterflies," 525.

76. Wallace, *The Geographical Distribution of Animals*, 552–53.

77. Browne, "A Science of Empire," 468; see also Zeller, "Classical Codes," 22.

78. Johnson, "Type-Specimens of Birds," 181.

79. Jardine, "Hints for Preparing and Transmitting Ornithological Specimens from Foreign Countries," 9.

80. Jardine, 9.

81. Raffles, *In Amazonia*, 133; Stagl, *A History of Curiosity*, 113.

82. Driver, *Geography Militant*, 49–67.

83. Adams, *Notes of a Naturalist*, 1.

84. Bree, *A History of the Birds of Europe*, 86. When volunteering in the Crimean War, Andrew Leith Adams misplaced his journal of his travels in Cashmere, which he regretted, "as it contained many valuable natural history notes. Adams, *Notes of a Naturalist*, 209.

85. Ó Cadhla, *Civilizing Ireland*, 106.

86. See Burant, "The Military Artist," 33–51; Burant, "Record of Empire," 120–28.

87. See Schwartz, "Photographs from the Edge of Empire," 154–71; Ryan, *Picturing Empire*.

88. Thomas Blakiston to Alfred Newton, from British North American Exploring Expedition, Saskatchewan River, BNA, 5 January 1858, Alfred Newton Papers, CUL MS Add.9839/1B/700, Cambridge University Library, Cambridge, UK.

89. Thomas Blakiston to Alfred Newton.

90. Thomas Blakiston to Alfred Newton.
91. Thomas Blakiston to Alfred Newton.
92. Raffles, "The Uses of Butterflies," 532.
93. Raffles, 532.
94. Camerini, "Evolution, Biogeography, and Maps," 701.
95. Bartholomew, *Bartholomew's Physical Atlas*. The atlas is "a series of maps illustrating the distribution of over seven hundred families, genera and species of existing animals, prepared by J. G. Bartholomew, W. Eagle Clarke and Percy H. Grimshaw; under the patronage of the Royal Geographical Society."

Chapter Two

1. Ranken, *Canada and the Crimea*, 297.
2. Rappaport, "Christmas in the Crimea." See Rupprecht, "Wonderful Adventures of Mrs. Seacole," 176–203.
3. Seacole, *Wonderful Adventures of Mrs. Seacole*, 187.
4. Blakiston, "Birds of the Crimea," 5603.
5. Smurthwaite, "Notes on the Great Bustard," 5517–18.
6. Blakiston, "Birds of the Crimea," 5603.
7. Blakiston, 5603.
8. Irby, "Lists of Birds Observed in the Crimea," 5359.
9. Irby, 5359.
10. Irby, *Ornithology of the Straits of Gibraltar* (1895), 253–57.
11. Irby, 254.
12. Kessler et al., "Satellite Telemetry Reveals Long-Distance Migration," 311–20.
13. Irby, *Ornithology of the Straits of Gibraltar* (1895), 256.
14. Stevenson, *The Birds of Norfolk*, 32, 36, 37, 39, 42.
15. British Museum of Natural History, *Guide to the Gallery of Birds*, 55.
16. Alonso, "The Great Bustard," 1–13; Burnside et al., "Human Disturbance and Conspecifics," 32–44.
17. Andryushchenko and Popenko, "Birds and Power Lines in Steppe Crimea," 34–41.
18. Alonso, "The Great Bustard," 1–13.
19. I am grateful to Paul Evans, librarian, at the Royal Artillery Museum, Firepower, at Woolwich, for his generous feedback on an earlier version of this chapter.
20. Blakiston, "Birds of the Crimea," 5348.
21. Dawson, *Soldier Heroes*, 9.
22. Woollacott, *Gender and Empire*, 61. For works on the politics of the Crimean War, see Conacher, *Britain and the Crimea*; Hearder, *Europe in the Nineteenth Century* 155–200; Goldfrank, *The Origins of the Crimean War*.
23. Ramsay, *The Ornithological Works of Arthur*, xxii, xxx.
24. Rupprecht, "Wonderful Adventures of Mrs. Seacole," 200.
25. Dawson, *Soldier Heroes*, 95.
26. Johnson, "Cast in Stone," 51–65.
27. Ramsay, *The Ornithological Works of Arthur*, xlii; Keene, "Hay, Arthur, Ninth Marquess of Tweeddale."

28. Rappaport, *Queen Victoria*, 294.

29. The Victoria and Albert Museum, London, houses a set of original albumen prints in the Prints and Drawings Study Room. Some original prints from Cundall and Howlett's series "Crimean Heroes and Trophies" are at the National Army Museum, London, England.

30. Anon., "Crimean Heroes and Trophies," 369.

31. Anon., "Russian Guns and Bells from Sebastopol," 209.

32. David Livingstone arrived at Southampton on 12 December 1856. Roberts, "Livingstone, David."

33. Green, *Dreams of Adventure, Deeds of Empire*, 8–13.

34. Green, 22–23, 58.

35. Boyd, "Manliness and the Mid-Victorian Temperament," 65.

36. Rawlinson, "The Military Advantages of a Daily Mail-Route to India through Turkey and the Persian Gulf," 181–91.

37. Anon., "To the Editor of the United Service Journal" (1830): 367–68.

38. Smyth, "To the Editor of the Journal." For works on the history of the United Service Institution, also known as the Royal United Service Institution (RUSI), see O'Connor, "The RUSI, Imperial Defence and the Expansion of Empire 1829–90."

39. Smyth, "To the Editor of the Journal."

40. Born in Gosport, Hampshire, he attended the Royal Military Academy at Woolwich and was commissioned as a second lieutenant, Royal Engineers, serving in Canada and Ireland with the Ordnance Survey. In 1851, Portlock was appointed inspector of studies at the Royal Military Academy, Woolwich. He was an ardent advocate of education in the army, especially in the scientific corps, and he instituted many valuable reforms in the system at Woolwich. His reforms were extended to the Military College, Sandhurst. Portlock, "On the Advantage of Cultivating the Natural and Experimental Sciences," 291; Vetch, "Portlock, Joseph Ellison."

41. Portlock, "On the Advantage of Cultivating the Natural and Experimental Sciences," 291.

42. Haley, *The Healthy Body and Victorian Culture*, 17–18; Portlock, "On the Advantage of Cultivating the Natural and Experimental Sciences, 290.

43. Portlock, "On the Advantage of Cultivating the Natural and Experimental Sciences," 306.

44. Ross, "The Cultivation of Scientific Knowledge," 774–81. Ross wrote his article before the Cardwell Reforms of 1870–71.

45. Browne, "A Science of Empire," 461.

46. Greer, "Placing Colonial Ornithology," 85–112.

47. Dawson, *Soldier Heroes*, 61, 82; Summers, "Pride and Prejudice," 33–56.

48. The presentation took place at the parade of the Horse Guards in London on 18 May 1855. McGilchrist, *The Public Life of Queen Victoria*, 180.

49. Blackwood, *A Narrative of Personal Experiences and Impressions*, 180.

50. Blackwood, 180. Blackwood was most likely alluding to Florence Nightingale, who used the old Barrack Hospital at Scutari as her nursing base during the Crimean War. For works on women in the Crimea, see Fletcher, "'Mother Seacole,'" 7–21.

51. The Blakiston family held an important position in the community of Lymington, with its family banner displayed prominently in the St. Thomas and All Saints Church. James, *Lymington*, 80–84; St. Thomas and All Saints, *St. Thomas and All Saints*, 10–13.

52. Hawker, *Instructions to Young Sportsmen*.

53. Blakiston, *Twenty Years in Retirement*, 129.

54. Blakiston, 159. John Blakiston was involved in the "Friends of the Reform" in the 1830s. Blakiston was born in the same year of the 1832 Reform Act, an issue that his family supported. James, *Lymington*, 86–87.

55. Burton and Peacey, *Soldiers of the Queen*, 19–21.

56. Johnson, "Richardson, Sir John"; Johnson and Johnson, "Richardson, Sir John."

57. Thomas W. Blakiston attended the second Royal Military Academy, Woolwich, situated on Woolwich Common (1806–1939).

58. Dickenson, *Drawn from Life*, 199; Smyth, *Sandhurst*, 54–55.

59. John Blakiston attended the original Royal Military Academy (1741–1806), Woolwich, within the walls of the Royal Arsenal. Blakiston, *Twelve Years' Military Adventure*, 11.

60. Carl Ritter was influenced by Immanuel Kant and moved away from dry facts, attempting to envision a more holistic view of the land by focusing on the living environments and their impact on humans. See Livingstone, *The Geographical Tradition*, 139–42.

61. Zeller, "Classical Codes," 22.

62. Anon., "Taxidermy," 424.

63. John Walter Wedderburn sent 152 specimens to Sir William Jardine from Halifax, Bermuda, and Surinam and published the article "Ornithology of Bermuda" when stationed there with his regiment. Jackson and Davis, *Sir William Jardine*, 173–74. Many officers contributed to the colonial meteorology of Halifax. See Morrison, "Soldiers, Storms and Seasons," 224–25.

64. Ranken, *Canada and the Crimea*, 4.

65. Edward Loftus Bland's North American birds are now housed at the National Ireland Museum, Natural History, in Dublin. Willis, "List of Birds of Nova Scotia," 280–86; Anon., "Museum and Library," 367.

66. Blakiston met Downs in Halifax, Nova Scotia, and continued to correspond with him, especially when he returned to British North America as part of the Palliser Expedition. Thomas Blakiston to Alfred Newton, 5 January 1858, from B.A. Exploring Expedition, Saskatchewan River, BNA, Newton Papers, University of Cambridge Archives, Cambridge, UK; Buggey, "Andrew Downs," 268–69.

67. Blakiston, "Birds of the Crimea," 5348.

68. Gupta and Ferguson, "Discipline and Practice," 13.

69. Dawson, *Soldier Heroes*, 55.

70. Mangan and McKenzie, "The Other Side of the Coin," 62–85.

71. Blakiston, "Birds of the Crimea," 5509.

72. Blakiston, 5674.

73. Blakiston, 5421.

74. Blakiston, 5352–53.

75. Livingstone, *Putting Science in Its Place*, 41.
76. Blakiston, "Birds of the Crimea," 5423.
77. Ramsay, *The Ornithological Works of Arthur*, xxxvi.
78. Ranken, *Canada and the Crimea*, 197.
79. Blakiston, "Birds of the Crimea," 5502.
80. Blakiston, 5423; Forbes, "William Yarrell, British Naturalist," 505–15.
81. Blakiston, 5503.
82. Blakiston, 5503.
83. Thomas Blakiston's youngest brother, Lawrence, died at the age of twenty at the Battle of the Redan on 8 September 1855.
84. Blakiston, "Birds of the Crimea," 5421.
85. Pratt, *Imperial Eyes*, 3.
86. Blakiston, "Birds of the Crimea," 5679.
87. Blakiston, 5511.
88. Anon., "Notices of Serials," 57.
89. Blakiston, "Birds of the Crimea," 5424.
90. Blakiston, 5349.
91. William Carte's birds of the Crimea are now housed at the Natural History Museum of Dublin. Sir James Alexander maintained an interest in birds throughout his military career and travels. He was also an RGS fellow and was sponsored by the Colonial Office and the RGS to explore southern Africa. Alexander, *Passages in the Life of a Soldier*, 278; Laidlaw, *Colonial Connections*, 33.
92. Blakiston, "Birds of the Crimea," 5512.
93. Irby, "Lists of Birds Observed in the Crimea," 5360.
94. See Burant, *Friendly Spies on the Northern Tour*.
95. Blakiston, "Birds of the Crimea," 5674.
96. Taylor, "Ornithological Observations," 234.
97. Taylor, 234.
98. Irby, "Lists of Birds Observed in the Crimea," 5360.
99. Blakiston, "Birds of the Crimea," 5362.
100. Pothunting was often associated with "peasant" and lower-class culture. For examples, see Boucher and Cruikshank, "Sportsmen and Pothunters," 1–18; Rome, "Nature Wars, Culture Wars," 432–53.
101. Blakiston, "Birds of the Crimea," 5511.
102. Ritvo, *The Platypus and the Mermaid*, 207.
103. In 1787, St. Petersburg ordered Russian troops to occupy the Crimean Peninsula, and the Crimean Tartar state was added to the Russian Empire in 1783 after Russian violated the Treaty of Küçük Kaynarca (1774) with the Ottoman Empire. Chirovsky, *An Introduction to Russian History*, 33, 44, 85.
104. The Russian Empire engaged in active efforts to annihilate the Crimean Tartar population commencing from the day of the "annexing" of the Crimea in 1783. Williams, *The Crimean Tatars*, 151.
105. Neilson also noticed: "The feathered tribe is numerous, comprising the crane, the heron, the blue crow, the beeeater, the hoopoo [sic], the starling, the oriol, the nightingale, and all the smaller birds we have in England; and the bright-coloured

plumage of the hoopoo, the oriol [sic], and the bee-eater, as they sit perched on the trees or flying in the sunshine, is really beautiful." Neilson, *Crimea*, 132, 135.

106. Irby, "Lists of Birds Observed in the Crimea," 5360.

107. Blakiston, "Birds of the Crimea," 5679. During the late eighteenth century, France valued the natural history collections of William V of Holland and therefore treated them as trophies of the war; specimens were transported by the tens of thousands to the museum in Paris (France invaded Holland in 1795). Pieters, "Notes on the Menagerie and Zoological Cabinet," 451–542.

108. Hawker, "Zoology from the Seat of War," 5203.

109. Blakiston, "Birds of the Crimea," 5349. Harriet Ritvo discusses the Zoological Society and its gardens at Regent's Park in *The Animal Estate*, 205–42.

110. Colonel Harding donated the specimen 12 September 1855. Sclater, *List of the Vertebrated Animals*, 293. Colonel Harding could possibly be Francis Pym Harding, a member of the Twenty-Second Regiment of Foot, the same regiment as Andrew Leith Adams, discussed in chapter 3. Harding served in the Crimea, Malta, and New Brunswick, Canada. Taylor, "Harding, Francis Pym."

111. Ritvo, *The Animal Estate*, 214–15.

112. Ritvo, 212, 206.

113. Carte, "Donations to the Royal Dublin Society," 286.

114. Blakiston, "Birds of the Crimea," 5679.

115. Blakiston, 5507.

116. Carte, "Report on the State and Progress," 40. For a work on science, empire, and Ireland, see Ó Cadhla, *Civilizing Ireland*.

117. Anon., *Penny Encyclopaedia of the Society for the Diffusion of Useful Knowledge*, 558–59; Anon., "Museum and Library," 441. The collection of the Royal Artillery Museum (founded in 1778) can be traced back to William Congreve, an officer of the Royal Artillery. George III, through the Board of Ordnance, requested that Congreve set up a collection for the betterment of Royal Artillery officers. The original collection was housed at the Royal Arsenal. In 1805, the collection was moved to the New Repository alongside the Royal Artillery barracks on the edge of Woolwich Common; in 1819, it was moved to the Rotunda building. General Sir John Lefroy (1817–90) was instrumental in building on the collections. In later years, Lefroy joined a scientific expedition to British North America and served as governor of Bermuda. The Rotunda collections were then transferred to the RAI in 1870. Timbers, *The Royal Artillery, Woolwich*, 120–21.

118. This information comes from a photograph titled *Crimean War Memorial at the Artillery Barracks, Woolwich*, 1860s, 182 by 233 mm, showing two soldiers standing at the base of the statue by John Bell; it is a female figure representing Liberty (or possibly Victory), nine feet high, on a sixteen-foot gray granite pedestal with four bronze shields. It was erected in 1860 to commemorate artillerymen killed in the Crimean War (1853–56). The print is stamped: "Photo. Estab. War Dept." The photograph is from the album *Malta, the Holy Land, etc*, 1860–1869, Y3011A/23, Royal Commonwealth Society Collections, Cambridge University Library, Cambridge, UK.

119. Wilford, "Proceedings of a General Meeting," 367. Henry Whitely Sr. was employed in Woolwich Arsenal, and was curator of the RAI. He had a natural history

agency in Wellington Street, Woolwich. British Museum, *The History of the Collections*, 510.

120. Blakiston's acquaintance with Newton might have been through William Samuel Newton, one of Alfred Newton's brothers, who served with the Coldstream Guards in the Crimean War from October 1854 to 8 April 1855. Blakiston mentioned in his letter to "remember me to your brother, when you write." Thomas Blakiston to Alfred Newton, 14 July 1859, Newton Papers, 9839/1B/701, University of Cambridge Archives, Cambridge, UK; Blakiston, "Interior of British North America," 141.

121. Royal Artillery Institution, *Minutes of Proceedings of the Royal Artillery Institution*, 55–58.

122. Royal Artillery Institution, 320.

123. Royal Artillery Institution, 320.

124. Thomas Blakiston to Alfred Newton, 14 July 1859, Newton Papers, 9839/1B/701, University of Cambridge Archives, Cambridge, UK.

125. Anon., "Review of Mr. Bree's 'Birds of Europe Not Observed in the British Isles,'" 88.

126. Blakiston, "Birds of the Crimea," 5674.

127. Blakiston, 5350.

128. Blakiston, 5422. For work on the rise of gentlemanly science and networks of trust, see Shapin, *A Social History of Truth*.

129. Blakiston, "Birds of the Crimea," 5679.

130. Newman was also author of *Birds-nesting* (1861), *New Edition of Montagu's Ornithological Dictionary* (1866), *Illustrated Natural History of British Moths* (1869), and *Illustrated Natural History of British Butterflies* (1871).

131. Pascoe, *The Hummingbird Cabinet*, 23.

132. Royal Artillery Institution, *Minutes of Proceedings of the Royal Artillery Institution*, 821–22.

133. Royal Artillery Institution, 821–22.

## Chapter Three

1. A version of this chapter was presented as Greer, "Birds and Biography," at the annual meeting of the Canadian Historical Association in 2010.

2. Adams, "Migrations of European Birds," 331.

3. Adams, 331.

4. Reichlin et al., "Migration Patterns," 393.

5. British Museum of Natural History, *Guide to the Gallery of Birds*, 173–74.

6. British Museum of Natural History, 173–74.

7. Drummond Hay, "Occurrence of the Hoopoe," cvlvii–cvlviii.

8. Drummond Hay, cvlvii–cvlviii.

9. Yarrell, *A History of British Birds*, 180.

10. Irby, *Ornithology of the Straits of Gibraltar*, 134–35.

11. Irby, 134–35.

12. Irby, 134–35.

13. Watkins, "The Ornithology of Andalusia," 5315.

14. Irby, *Ornithology of the Straits of Gibraltar*, 134–35.
15. Kristin, "Family Upupidae (Hoopoes)," 396–411.
16. Irby, *Ornithology of the Straits of Gibraltar*, 134–35.
17. Taylor, "Ornithological Observations," 230.
18. White, *The Natural History of Selborne*, 25.
19. Adams, *Notes of a Naturalist*, 7–8.
20. Irby, *Ornithology of the Straits of Gibraltar*, 134–35.
21. Irby, 68.
22. Kunstmann, The *Hoopoe*, 13–18.
23. The species was mentioned in *The Historical Atlas of Breeding Birds in Britain and Ireland 1875-1900*, which cited Jardine in *The Naturalist's Library*, 381.
24. Drummond Hay, "Occurrence of the Hoopoe," cvlvii–cvlviii.
25. BirdLife International, "*Upupa epops*."
26. Kristin, "Family Upupidae (Hoopoes)," 396–411.
27. Morris, "Rare Exotic Hoopoe Bird."
28. Adams, *Notes of a Naturalist*, 75.
29. Adams, 75.
30. Adams, 75.
31. Adams, 75–76.
32. Adams, viii; Gaston, "Adams, Andrew Leith."
33. Adams, x.
34. Adams and Adams, *On Ornithology as a Branch of Liberal Education*, 29, 33.
35. By the mid-nineteenth century, the British army moved away from practices of corporeal punishment to ways of improving the efficiency of military manpower through other forms of disciplinary power. Many of these initiatives were part of the Cardwell Reforms introduced by the British army under the Gladstone government between 1868 and 1874. Andrew Leith Adams published many articles on military fitness and recruiting, which were listed in his obituary (Anon., "Obituary: Andrew Leith Adams," 338). These works included "The Recruiting Question from Medical and Military Points of View," *Journal of the United Service Institution*; "Heredity of Abnormalities and Deformities," *The Lancet*; and "On the Physical Requirements of the Soldier," *British and Foreign Medical Chirurgical Review*. Rodger, *Aberdeen Doctors*, 329.
36. Definitions of temperate and tropical climatic zones have been at the center of geographic debates: Bailey, "Toward a Unified Concept of the Temperate Climate," 516–45; Okihiro, "Unsettling the Imperial Sciences," 745–58. For Aristotelian ideas on temperate and temperance, see Young, "Aristotle on Temperance," 521–42.
37. Duncan, "The Struggle to Be Temperate," 34–47.
38. Duncan, 40.
39. Rose, *Which People's War?*
40. Curtin, "Disease and Imperialism," 102; Bewell, *Romanticism and Colonial Disease*, 279. Other works on the effects of climate on the imperial British body include Arnold, *Colonizing the Body*; Harrison, *Public Health in British India*; Peers, "Soldiers, Surgeons," 137–60.
41. James Hunt quoted the work of French army officer, medical geographer, and ethnologist Jean Christian Boudin. Hunt, "On Ethno-climatology," 136.

42. Ogborn and Philo, "Soldiers, Sailors and Moral Locations," 221.

43. Ogborn and Philo, 223.

44. Adams, *Field and Forest Rambles*, 43.

45. Adams, 43.

46. For works on the links between natural history and civic science, see Finnegan, "'An Aid to Mental Health,'" 326–37; Finnegan, *Natural History Societies*.

47. Livingstone, *Science, Space, and Hermeneutics*, 63.

48. The region would later be known as the Royal Deeside when Queen Victoria purchased Balmoral Castle.

49. Adams and Adams, *On Ornithology as a Branch of Liberal Education*, v.

50. Adams and Adams, vi.

51. Adams and Adams, vii–viii, 12.

52. Francis Adams was born at Auchinhove, Lumphanan, and attended Kings College, Aberdeen, where he studied classics. After graduating in 1813, he pursued medicine at the University of Edinburgh and became a member of the Royal College of Surgeons in London in December 1815. He was later offered the chair of Greek studies at the University of Aberdeen. Ford, "Francis Adams," 56; Adams, *The Genuine Works of Hippocrates*.

53. Withers, *Geography, Science and National Identity*, 159.

54. Withers, 154; Hargreaves, *Aberdeenshire to Africa*, 2.

55. Dr. Robert Wilson encouraged students to take sketching and photographic materials with them, as well as scientific instruments and a diary to record their observations. Wilson served with the East India Civil Service and was private secretary to the Marquis of Hastings and governor of Malta. The Marischal Museum houses a variety of Wilson's traveling artifacts, including his botanical microscope (ABDUA 37170), his sporting gun (ABDUA 36838), and a shrub specimen from Gozo, Malta (ABDUA 63382). Rodger, *Aberdeen Doctors*, 306.

56. Denny, "British Temperance Reformers," 333.

57. Denny, 335.

58. Denny, 329.

59. A detailed version of MacGillivray's life was written by another William MacGillivray, whose book was published forty-nine years after the ornithologist MacGillivray's death. McGillivray, *A Memorial Tribute to William McGillivray*, 59.

60. Army Medical Department, *A Catalogue of the Collection of Mammalia and Birds*, iii; Mackenzie, *Museums and Empire*, 80.

61. Sir James McGrigor (1771–1858) had studied at the college and helped to shape the Royal Army Medical Corps. See McGrigor, *The Scalpel and the Sword*; Withers, *Geography, Science and National Identity*, 159.

62. The catalog included a list of all of the species collected from across the British Empire. Army Medical Department, *A Catalogue of the Collection of Mammalia and Birds*.

63. For example, the collection was used by Sir John Richardson and William Swainson in their *Fauna Boreali-Americana*, lxiii.

64. Andrew Leith Adams to John Gould, 19 January 1860, Gould Correspondence, Adams A L 15, Natural History Museum Library and Archives, London, UK.

65. Livingstone, *The Geographical Tradition*, 233–35.

66. Anon., "Obituary: Andrew Leith Adams," 338. The Second Anglo-Sikh War was a conflict between the British East India Company and the Sikh Empire, which resulted in the subjugation of the Sikh Empire and the annexation of the Punjab. This territory would later become the North-West Frontier Province.

67. David N. Livingstone has stated that "seasoning," "acclimation," and "acclimatization" meant the same thing at this time. See Livingstone, "Tropical Climate and Moral Hygiene," 101.

68. Early ideas on acclimatization were based on James Johnson's *The Influence of Tropical Climates on European Constitutions* (1813). See Bewell, *Romanticism and Colonial Disease*, 279–80.

69. Arnold, *The Tropics and the Traveling Gaze*, 15.

70. Gaston, "Adams, Andrew Leith." Adams also served in medical charge of the detachment of the Twenty-Second Regiment of Foot, with the expeditionary force under Sir Sydney Cotton, in 1854, against the Mohmund nation on the Peshawur frontier. Anon., "Obituary: Andrew Leith Adams," 338.

71. Adams collected the type specimens of the orange bullfinch (*Pyrrhula aurantiaca*) and the black-winged snowfinch (*Montifringilla adamsi*) now housed at the British Museum. The British Museum also houses Adams's Kashmir martin (*Chelidon cashmeriensis*) and the white-breasted Asiatic dipper (*Cinclus cashmeriensis*), which were donated by John Gould. The Natural History Museum of Ireland houses Adams's solitary snipe (*Gallinago solitaria*) from the Western Himalayas. See Adams, "The Birds of Cashmere and Ladakh," 169–90; Adams, "Notes on the Habits, Haunts, Etc.," 466–512; Adams, *Wanderings of a Naturalist in India*, 97; Gaston, "Adams, Andrew Leith."

72. See Stanley, *White Mutiny*.

73. Adams, *Wanderings of a Naturalist in India*, 245.

74. Adams, 245. Adams most likely learned of Humboldt from William MacGillivray, who translated Humboldt's *Travels and Researches of Alexander von Humboldt* (1832).

75. Arnold, *The Tropics and the Traveling Gaze*, 111–14.

76. Adams, *Wanderings of a Naturalist in India*, 206.

77. Adams, *Notes of a Naturalist*, 52.

78. Adams, *Wanderings of a Naturalist in India*, 150.

79. British anxieties about hot climates have been traced back to their earliest encounters with tropical environments. See Kupperman, "Fear of Hot Climates," 213–40.

80. Adams, *Wanderings of a Naturalist in India*, 10.

81. Adams, *Notes of a Naturalist*, ix.

82. Edwards, Gosden, and Phillips, *Sensible Objects*, 7.

83. Adams and Adams, *On Ornithology as a Branch of Liberal Education*, vii.

84. Latin for "the land of one's birth."

85. Adams and Adams, *On Ornithology as a Branch of Liberal Education*, vi.

86. Adams and Adams, vi.

87. Adams, *Wanderings of a Naturalist in India*, 137. His residence was in Poonah (now Pune).

88. Adams, 137.

89. Kennedy, *The Magic Mountains*, 118.

90. Bird, "The Acclimation of European Troops," 324. The system of hill stations in India also helped to act as "a barrier again invasion from the north." Clarke, "The Military Advantages of a Daily Mail Route," 181.

91. Adams, *Wanderings of a Naturalist in India*, 53–54.

92. Adams, 30.

93. Adams, *Notes of a Naturalist*, 75.

94. Three years after Prime Minister Benjamin Disraeli purchased shares of the Suez Canal, he was able to bring 7,000 troops to Malta, with great speed and secrecy, at a time when he wished to make a military-diplomatic gesture to bring pressure on Russia. See Garratt, *Gibraltar and the Mediterranean*, 137.

95. Livingstone, *Science, Space and Hermeneutics*, 49.

96. Livingstone, 61. For examples specific to Malta, see Jankovic, "The Last Resort," 285.

97. Baynes, "Malta," 340; Dr. John Davy, *Report on the Disease etc. of the Garrison of Malta, for the Year Ending the 20th Dec 1828*, 4, R1MSJD77, Royal Institution, London, UK. Dr. John Davy served as a military surgeon and maintained an interest in birds. The Davy materials housed at the Royal Institution include "Lecture 34: Of Birds in General," lecture notes that described the migratory behaviors of birds. John Davy Notebook, 1802–1843 R1MSJD 1/2, Royal Institution, London, UK.

98. Adams, "Notes on Certain Meteorological Phenomena," 306.

99. Adams, 306.

100. Adams, *Notes of a Naturalist*, 97–98.

101. The exact date of the establishment of the Malta Garrison Library is still uncertain. However, Giuseppe Pericciuoli Borzesi mentions the library as early as 1830 in *The Historical Guide to the Island of Malta*, 40.

102. William Reid was born at Kinglassie, Fife, in Scotland, attended the Royal Military Academy, Woolwich, and served in the Peninsular Wars. Reid served as governor of Bermuda (1839–46) and the British Windward Islands (1846–1848), and contributed to meteorology and published many works, including *An Attempt to Develop the Law of Storms by Means of Facts* (1838; 3rd ed., 1850) and *The Progress of the Development of the Law of Storms and of the Variable Winds* (1849). Reid became a fellow of the Royal Society in February 1839. See Blouet, "Sir William Reid," 169–91.

103. Blouet, "Sir William Reid." A general library was set up in 1825 in Malta by the Eightieth Regiment of Foot. See Vickers, "A Gift So Graciously Bestowed," 10.

104. Denny, "British Temperance Reformers," 329–45.

105. Malta Garrison Library, *Second Part of the Classified Catalogue*, 123.

106. Peers, "Privates off Parade," 839.

107. Charles Wright wrote in his notebook on 24 March 1862, "Made the acquaintance of Mr. A. R. Wallace, a distinguished Naturalist." Wallace showed him "two male Birds of Paradise from New Guinea and some rare species of parrots, he brought with him from the Malay Archipelago, where he has been collecting specimens of birds and insects for several years past." Wallace purchased a specimen of Hoopoe in the market. In Charles Wright, *Rough Notes—Birds of Malta Oct 1861 to September 1862 vol IV*, 24 March 1862, Natural History Museum, Tring, UK.

108. Frome, "Moncrieff's System of Artillery," 33.

109. For a detailed history on Malta and the Grand Tour, see Freller, *Malta and the Grand Tour*.

110. By the mid-nineteenth century, Britain viewed Malta as unhealthy because of the cholera epidemics. Jankovic, "The Last Resort," 271–98; Pemble, *The Mediterranean Passion*, 150.

111. Freller, *Malta and the Grand Tour*, 172.

112. Hannah, *Governmentality and the Mastery of Territory*, 20.

113. Colley, *Britons*, 273; Kennedy, *The Magic Mountains*, 147.

114. Charles Wright "made the acquaintance of a lady, Mrs. Blackburn, who is on her way to the Nile, and is very much interested in birds." Wright, *Rough Notes—Birds of Malta 1861 to September 1862 vol IV*, 24 March 1862; Adams, *Notes of a Naturalist*, 228–29.

115. Andrew Leith and Bertha Adams's firstborn son, Francis Adams, was born in Floriana, Malta, 27 September 1862; Tasker, *"Struggle and Storm,"* 17–20.

116. Anon., "How I Spent My Five Weeks' Leave," 427.

117. Adams, *Notes of a Naturalist*, 76.

118. Adams, 75.

119. Adams, 97–98.

120. Adams, 97–98.

121. Adams, 90.

122. Adams, 90.

123. Moore, Pandian, and Kosek, "The Cultural Politics of Race and Nature," 12.

124. Wright, "List of the Birds," 44.

125. Adams, *Notes of a Naturalist*, 88.

126. Adams, 88.

127. Scicluna and Knepper, "Prisoners of the Sun," 502–21.

128. Britain took Malta from the French in 1800 and maintained possession in 1802 with the Treaty of Amiens during the Napoleonic Wars. In 1814, European powers recognized Malta as a British Crown colony with the signing of the Treaty of Paris. Martin, *History of the Colonies of the British Empire*, 152.

129. Hugh E. Strickland, "On the Occurrence of *Charadrius Virginiacus*," 40. Strickland examined the collection of Lieutenant Colonel Henry Maurice Drummond Hay, Forty-Second Regiment (Black Watch), who collected birds in the Mediterranean stations of the Ionian Islands and Malta in the late 1840s. Drummond Hay would be the first president of the BOU.

130. Bourguet et al., *L'invention scientifique de la Méditerranée*, 97–116.

131. Maempel, "T. A. B. Spratt," 271–308.

132. Thomson, "Notice of Migratory Birds," 125.

133. Wright, "List of the Birds," 44; Adams, "Migrations of European Birds," 325.

134. Adams, "Migrations of European Birds," 325.

135. Adams befriended Charles Wright, the editor of the *Malta Times*. Wright published his list in the January issue of the *Ibis* in 1864. Adams, *Notes of a Naturalist*, 95; Crispo-Barbaro, *The Birds of Malta*, 5.

136. Wright, "List of the Birds," 42.

137. Adams, "Migrations of European Birds," 324. Andrew Leith Adams helped to establish the Society of Archaeology, History, and Natural Sciences in Malta, 1866. Maempel, *Pioneers of Maltese Geology*, 233.

138. For details on Adams's geologic work in Malta, see Maempel, *Pioneers of Maltese Geology*.

139. Charles Wright, *Rough Notes—Birds of Malta Oct 1861 to September 1862 vol IV*, 29 November 1861, Natural History Museum, Tring, UK.

140. Adams, *Notes of a Naturalist*, 337.

141. Playfair, *Handbook to the Mediterranean*, 175.

142. Darwin, "Imperialism and the Victorians," 614-42.

143. Ramm, "Great Britain and France in Egypt," 73-119.

144. Charles Wright documented his trip with Andrew Leith Adams to Tunis in his field journal on 6 April 1858. They were granted passage in HMS *Wanderer* and stayed in Tunis from 4 March to 14 March. Charles Wright, *Rough Notes on Birds of Malta Aug 1855 to May 1858 vol 1*, 6 April 1858.

145. England has had a long history of trading with Tunis, starting in 1551, and more formally with the first English consul stationed there in 1558. Trade between Tunis and Malta declined when North Africa was portioned by Italy and France. Manai, "Anglo-Tunisian Commercial Relations," 365-72.

146. Carriage, "France, England, and the Tunisian Affair," 35.

147. Manai, "Anglo-Tunisian Commercial Relations," 367-68.

148. Adams authored "Malta: Cholera Epidemic, Importance of Hygiene; Prepared a Report on the Epidemic" with Assistant-Surgeon F. H. Welch. Anon., "Obituary: Andrew Leith Adams," 338. Adams traveled with Scottish lawyer and Egyptologist Alexander Henry Rhind (1833-63).

149. On the geographic imagination of Egypt, see Gregory, "Between the Book and the Lamp," 29-57; Gregory, "Emperors of the Gaze"; Jasanoff, *Edge of Empire*.

150. Shelley, *A Handbook to the Birds of Egypt*, 87.

151. Adams, *Notes of a Naturalist*, 14.

152. Adams, 14.

153. Adams donated thirty-three birds from "N.E. Africa" to the British Museum in 1864. British Museum, *The History of the Collections*, 293. Early nineteenth-century French engagements in Egypt, in particular, attempted to annex Egypt as part of the Europe continent, and an integral and historic site in maintaining power in the region. See Said, *Orientalism*, 84-85.

154. Henry Baker Tristram (1822-1906) served as army chaplain in Bermuda and would later become the canon of Durham Cathedral. He traveled to Malta, North Africa, Israel, Palestine, and Syria and made a collection of birds. He was also one of the founding members of the BOU. He published several books, including *The Great Sahara* (1860), *The Land of Israel, a Journal of Travels with Reference to Its Physical History* (1865), and *The Natural History of the Bible* (1867). Many of his bird skins are housed at the World Liverpool Museum, Liverpool.

155. Wollaston, *Life of Alfred Newton*, 25.

156. Adams, *Field and Forest Rambles*, 1.

157. Adams, 1.

158. Adams, 12. The First Battalion of the Twenty-Second Regiment replaced the Fifteenth Regiment of Foot, which served in New Brunswick for five years and had orders to rotate to Bermuda in the third week of April. Dallison, *Turning Back the Fenians*, 85.

159. Dallison, 85.

160. Dallison, 85.

161. Taylor, "Harding, Francis Pym"; Jones and Jones, *Fredericton and Its People*, 5. Their first son, Francis Adams, born in Malta, lived in New Brunswick and became a well-known author as well. Casey, "Adams, Bertha Jane Leith"; Tasker, "Adams, Francis William Lauderdale."

162. Numerous British military officers contributed to the natural history of Nova Scotia, including Captain Thomas Wright Blakiston (Royal Artillery) and Edward Loftus Bland (Royal Engineers).

163. McGhie, *Henry Dresser and Victorian Ornithology*. The Manchester Museum houses several bird skins collected by Henry Dresser and his brother, Arthur, from New Brunswick. Arthur Dresser Fonds, "Travels in New Brunswick, Canada and Manitoba, 1868–1877," 5, A-1536, Library and Archives Canada, Ottawa. Arthur included a "List of bird skins" from Musquash, 10–21.

164. Gordon, "Wilderness Journeys in New Brunswick," 457.

165. Jasen, *Wild Things*; Joan M. Schwartz, "William Notman's Hunting Photographs," 20–29; Parenteau, "Angling, Hunting and the Development of Tourism," 10–19.

166. Adams, *Field and Forest Rambles*, 3.

167. Adams, 38.

168. Adams, 17, 21.

169. Adams, 17.

170. Adams, 193.

171. Adams, 144.

172. Adams, 39.

173. Adams, 121.

174. Adams, 121. During his station in New Brunswick, Adams visited George Augustus Boardman's collection, which numbered over several thousand bird specimens. Boardman was well connected to an American network of ornithologists and museums, including the Smithsonian Institution and the Boston Natural History Society. Many colonial officials visited Boardman's collection of birds, including Sir Arthur Gordon, as well as fellow ornithologists such as Henry Dresser, who fostered a relationship with Boardman over the years. Boardman developed a serious interest in birds after conducting a business trip to the Caribbean region in 1840–41. McAlpine, "Boardman, George Augustus"; McGhie, *Henry Dresser and Victorian Ornithology*. Adams mentions Boardman in his book *Field and Forest Rambles*, 14–15.

175. Adams, *Field and Forest Rambles*, 1.

176. Adams, 285.

177. Adams, 285.

178. Adams, 184.

179. Bird, *North Shore New Brunswick Regiment*, 69.

180. Lambert and Lester, *Colonial Lives across the British Empire*, 2.
181. Adams, *Notes of a Naturalist*, 75.
182. Adams, *Field and Forest Rambles*, 43.
183. Gaston, "Adams, Andrew Leith."
184. Anon., "Obituary: Andrew Leith Adams," 338. After reviewing his life geography, one might be apt to also credit more than a few birds.

## Chapter Four

1. Irby, *Ornithology of the Straits of Gibraltar*, 76–77. Irby's book was first published in 1875.
2. Irby, 76–77.
3. Some of the earliest specimens of golden oriole sent to Britain were from Gibraltar. The English naturalist George Edwards had one sent to him that was shot on the Rock of Gibraltar in the mid-eighteenth century. See Yarrell, *A History of British Birds*, 231; Irby, *Ornithology of the Straits of Gibraltar*, 76–77.
4. BirdLife International, "Oriolus oriolus."
5. Irby, *Ornithology of the Straits of Gibraltar*, 76–77.
6. Adams, *Notes of a Naturalist*, 103.
7. Adams, 103.
8. Taylor, "Ornithological Observations," 229.
9. Jardine, *The Naturalist's Library*, 103.
10. Batten, Bibby, and Clement, *Red Data Birds in Britain*, 263–64.
11. Yarrell, Newton, and Saunders, *A History of British Birds*, 235–36.
12. Yarrell, Newton, and Saunders, 235, 238.
13. British Museum of Natural History, *Guide to the Gallery of Birds*, 133.
14. Yarrell, Newton, and Saunders, *A History of British Birds*, 236.
15. Maxwell, "Birds," 925.
16. BirdLife International, "Oriolus oriolus."
17. Anon., "New Protection for Migratory Birds."
18. Irby's book was first published in 1875. Irby, *Ornithology of the Straits of Gibraltar*, 2–3. Born at Boyland Hall in Morningthorpe, Norfolk County, Irby descended from a family steeped in the British army and the Royal Navy. His father, Frederick Paul Irby (1779–1844), served on the British and North American stations and patrolled the west coast of Africa as part of the abolitionist movement. His uncle Charles Leonard Irby (1789–1845) spent time in the Mediterranean, Cape of Good Hope, Montevideo, Mauritius, and North America, publishing his travel experiences. Another uncle, Edward Methuen Irby (1788–1809), served with the Third Regiment of Foot and was killed in action during the Peninsular Wars. Irby's father served with the Royal Navy and enforced antislavery laws when stationed on the HMS *Amelia* off the West African coast from 1811 to 1813. During his own tour of duty, Captain Irby rescued three child slaves, who in 1813 were baptized at Saint Peter Mancroft Church in Norwich.
19. Irby, 2–3.
20. Irby served with the Ninetieth Regiment in the Crimea and in India. Anon., "Irby's Birds of Gibraltar," 364.

21. Works by British military officers L. Howard Irby and Willoughby Verner continue to be used as a baseline for comparing breeding bird populations and migration patterns. See Finlayson, *Birds of the Strait of Gibraltar*, xxiii; Lathbury, "A Review of the Birds of Gibraltar," 25.

22. In classical times, Gibraltar was known as Mons Calpe and together with Mons Abyla on the African coast formed the great Pillars of Briareus to the Greeks and the Pillars of Hercules to the Romans. Dutch and British interests in Gibraltar converged in 1704, when it fell to the British during the War of the Spanish Succession. Gibraltar was viewed by European powers as a strategic location for the trade of commodities to Asia. Although 4,000 Spanish inhabitants fled the region, some stayed with Jewish trading groups and traders from Genoa, Malta, and Britain to settle the British "trading" outpost. Gibraltar was formally ceded to Britain by Spain in the 1713 Treaty of Utrecht and was only formally declared a colony in 1830. Garratt, *Gibraltar and the Mediterranean*, 11–13; Skinner, "British Constructions with Constitutions," 301–20.

23. For an example, see Schwartz and Ryan, *Picturing Place*.

24. Paasi, "Territorial Identities as Social Constructs," 93. For studies specific to Gibraltar, see Lambert, "'As Solid as the Rock?,'" 206–20; Dodds, Lambert, and Robison, "Loyalty and Royalty," 365–90.

25. Foucault, *Discipline and Punish*, 164, 135.

26. Myerly, *British Military Spectacle*, 8–9; Schieffelin, "Problematizing Performance," 195.

27. Ó Cadhla, *Civilizing Ireland*, 136.

28. Bartlett, *Gleanings on the Overland Route*, 158.

29. Richardson, *The Anglo-Indian Passage*, vi. The nineteenth century has been described as a significant period in altering ideas of space and time. The term "time-space compression" is used in geography to describe the processes that tend to accelerate the experience of time and reduce the significance of distance during a particular historical moment. See May and Thrift, *TimeSpace*, 7; Harvey, *The Condition of Postmodernity*, 240.

30. Richardson, *The Anglo-Indian Passage*, vi.

31. Irby, "Notes of Birds," 221. See Collingham, *Imperial Bodies*. See also Bayly, *Empire and Information*. Bayly's work provides an examination of the ways in which relations between local informants and British officials changed with the advent of the Indian Mutiny. Hearsay knowledge was replaced with statistics and numerical data.

32. Collingham, *Imperial Bodies*, 2.

33. Graeme Wynn, foreword to Gillespie, *Hunting for Empire*, xix.

34. Nash, "Performativity in Practice," 653–64.

35. Dawson, *Soldier Heroes*, 83.

36. The creation of the "Gibraltar tradition" was an important aspect of the mythology of the British Empire. However, a number of British statesmen critiqued British foreign policy of Gibraltar by the mid-nineteenth century, which caused an increased emphasis on illustrating Gibraltar's importance to Britain and the British Empire. Garratt, *Gibraltar and the Mediterranean*, 127–29. See also Lambert, "'As Solid as the Rock?,'" 206–20; Constantine, "Monarchy and Constructing Identity," 23–44.

37. Boulton, *Reminiscences of the North-West Rebellions*, 22.

38. Anon., "Notes on Gibraltar," 51.
39. Martin, *History of the British Possessions*, 110.
40. Power, *Recollections of a Three Years' Residence in China*, 44.
41. Power, 2.
42. Power, 2.
43. Sayer, *The History of Gibraltar*, 481.
44. Gilbard, *A Popular History of Gibraltar*, 17; Power, *Recollections of a Three Years' Residence in China*, 2; Sayer, *The History of Gibraltar*, 481.
45. Sayer, *The History of Gibraltar*, 481.
46. Howe, "Mapping a Sacred Geography," 231.
47. Gilbard, *A Popular History of Gibraltar*, 22–23.
48. Napier, *Wild Sports in Europe, Asia and Africa*, 48. Napier, the adopted son of Sir Charles Napier, served at Gibraltar starting in 1837. Chichester, "Napier, Edward Delaval Hungerford Elers."
49. Drinkwater, *A History of the Siege of Gibraltar*, 282; Bartlett, *Gleanings on the Overland Route*, 168. See also Anon., *Gibraltar and Its Sieges*, 84.
50. Gilbard, *A Popular History of Gibraltar*, 22.
51. Boulton, *Reminiscences of the North-West Rebellions*, 19.
52. Headley, *The Life and Travels of General Grant*, 271–72.
53. Anon., "Gibraltar and Neighbourhood," 711.
54. Martin, *History of the British Possessions*, 110.
55. In 1889, an ordinance was issued by the governor of Gibraltar which decreed that only native-born inhabitants had the right of residence in the colony. Everyone else, including British subjects (exempting Crown officials), had to obtain permission to live on the Rock. Skinner, "British Constructions with Constitutions," 301–20; Burke and Sawchuk, "Alien Encounters," 531–61; Perera, "The Language of Exclusion," 209–34.
56. Coleridge in *Rock of Empire*, 57.
57. Thackeray in *Rock of Empire*, 76.
58. Finlayson, *Gibraltar: 300 Years of Images*, 126–27.
59. When at Gibraltar, Pindar stayed aboard HMS *Simoom*, the same troopship that Andrew Leith Adams took from Malta to New Brunswick. Pindar, *Autobiography of a Private Soldier*, 94.
60. Pindar, 94–95.
61. Archer, *Gibraltar, Identity and Empire*, 45.
62. Lambert, "'As Solid as the Rock?,'" 212.
63. Sanchez, *The Prostitutes of Serruya's Lane*, 54; Sala, *From Waterloo to the Peninsula*, 247.
64. Sanchez, *The Prostitutes of Serruya's Lane*, 45.
65. Anon., "Irby's Birds of Gibraltar," 364.
66. Anon., 364.
67. Irby might have attended the college with Lieutenant Colonel Henry Haversham Godwin-Austen (1834–1923), a British military officer with the Twenty-Fourth Regiment of Foot, who contributed to field ornithology in India.
68. Irby, *Ornithology of the Straits of Gibraltar*, 5.
69. Wolseley, *Soldier's Pocket-Book for Field Service*, 165.

70. Verner, "Obituary," 502; Wolseley, *Soldier's Pocket-Book for Field Service*, 179.

71. Verner, *My Life among the Wild Birds of Spain*, 39; Finlayson, "William Willoughby Cole Verner," 91–99.

72. Willoughby Verner spent many years at Gibraltar with the Rifle Brigade and eventually retired there. He was also professor of topography at Sandhurst Royal Military College and wrote numerous works on his activities in Africa.

73. See page 28 of Philip Savile Grey Reid, *MSS Stray Notes on Ornithology, 1871–1890*, Philip Savile Grey Manuscript Collection, Natural History Museum Ornithological Library, Tring, UK.

74. Ogborn and Philo, "Soldiers, Sailors and Moral Locations," 221.

75. Becher, "Zoological Notes from Gibraltar," 100.

76. See page 3 of Reid's *MSS Stray Notes on Ornithology*. See chapter 5.

77. Irby, *Ornithology of the Straits of Gibraltar*, 36.

78. The popular red-light district in nineteenth-century Gibraltar was called Serruya's Lane.

79. Pindar, *Autobiography of a Private Soldier*, 96.

80. Pindar, 96.

81. Anon., *United Service Magazine*, 105.

82. I viewed these books during my visit to the Gibraltar Garrison Library in February 2008. Gibraltar Garrison Library, *Catalogue of the Books in the Gibraltar Garrison Library*.

83. Robertson, *Journal of a Clergyman*, 201.

84. Irby, "Notes on the Birds," 343.

85. Verner, "Obituary," 504; Irby, *Ornithology of the Straits of Gibraltar*, 173.

86. Irby, *Ornithology of the Straits of Gibraltar*, 33.

87. Irby, 133.

88. Irby, 133. Interestingly, the millinery industry supplied feathers to some units of the British army. See Moore-Colyer, "Feathered Women and Persecuted Birds," 57–73.

89. Irby, *Ornithology of the Straits of Gibraltar*, 100; Reid, "Winter Notes from Morocco," 245.

90. Irby, *Ornithology of the Straits of Gibraltar*, 78.

91. Napier, *Wild Sports in Europe, Asia and Africa*, 52.

92. Reid, *MSS Stray Notes on Ornithology*, 2.

93. Pratt, *Imperial Eyes*, 8–9.

94. Garratt, *Gibraltar and the Mediterranean*, 129.

95. Irby, *Ornithology of the Straits of Gibraltar*, 26.

96. Irby, 3.

97. Watkins, "The Ornithology of Andalusia," 5312.

98. Irby, *Ornithology of the Straits of Gibraltar*, 31.

99. Irby, 68.

100. Reid described this particular collector, Captain Arthur Cowell Stark, during his meeting with Irby at the Zoological Gardens at Regent's Park, London, 22 April 1877, in his "Stray Notes on Ornithology, 1871–1890." Stark (1846–99) was a British physician and traveling naturalist who coauthored "Fauna of South Africa." He was killed during the Boer War.

101. Verner, *My Life among the Wild Birds of Spain*, 58–59.

102. Allen, "Young England," 114–32.

103. Phillips, *Mapping Men and Empire*, 14.

104. Drummond-Hay fostered special relations with Sultan Moulay Abderahman to maintain territorial influence in the region. Srhir, Williams, and Waterson, *Britain and Morocco*, 36; Archer, *Gibraltar, Identity and Empire*; Erzini, "Hal Yaslah Li-Taqansut (Is He Suitable for Consulship?)," 517–29.

105. Irby, *Ornithology of the Straits of Gibraltar*, 6.

106. Irby, 13.

107. According to Cynthia J. Becker, some Amazigh groups in Morocco do not eat tame or wild birds, as they believe them to be the souls of the dead. Becker, *Amazigh Arts in Morocco*, 208.

108. Irby, *Ornithology of the Straits of Gibraltar*, 208.

109. Irby, 6–7. According to Linda Colley, Tangier has been a reminder of "the importance of the Mediterranean as a cockpit for contending states and religions, as a place of commerce, and as a site of empire." See Colley, *Captives*, 33.

110. John Drummond Hay to the Earl of Derby, Zoological Society of London, 13 September 1832 and 5 July 1832, Library of the Zoological Society of London, London, UK.

111. John Drummond Hay to the Earl of Derby.

112. John Drummond Hay to the Earl of Derby.

113. Anon., "Additions to the Zoological Society's Gardens," 371.

114. Irby obtained the MSS and donated it to the Zoological Museum of the University of Cambridge. Anon., "Irby's Birds of Gibraltar," 364.

115. Irby, *Ornithology of the Straits of Gibraltar*, 2.

116. Irby, 7.

117. Irby, 10.

118. Irby, 8.

119. Philip Savile Grey Reid, "A Trip into Western Barbary," 15, private collection of Dave Reid.

120. Dalton, "The Colonial Conference 1887," 32–33.

## Chapter Five

1. Snow and Perrins, *The Birds of the Western Palearctic*; Ferguson-Lees and Christie, *Raptors of the World*; BirdLife International, "Pandion haliaetus."

2. Feilden, "Ospreys in Hampshire," 2991–92.

3. British Museum of Natural History, *Guide to the Gallery of Birds*, 80.

4. Adams, *Notes of a Naturalist*, 280.

5. Wright, "List of the Birds," 45.

6. Wright, 45.

7. Adams, *Notes of a Naturalist*, 21.

8. Irby, "Notes on the Birds," 222.

9. Irby, *Ornithology of the Straits of Gibraltar*, 197–98.

10. Reid, "The Birds of the Bermudas," 26.

11. The ospreys in North America migrate to South America. Alerstam, Hake, and Kjellén, "Temporal and Spatial Patterns," 555-66.

12. Holling and the Rare Breeding Birds Panel, "Rare Breeding Birds," 352-416.

13. RSPB, "Ospreys in the UK."

14. BBC, "Ospreys Return to Northumberland from Africa."

15. BirdLife International, "*Pandion haliaetus*."

16. Born in Argentina, W. H. Hudson (1841-1922) pursued field ornithology in Hampshire County. One of his most important works was *Hampshire Days* (1903). See Tippett, "W. H. Hudson in Hampshire."

17. Gould, *An Introduction to the Birds of Great Britain*, iii.

18. Matless, "Moral Geographies of English landscape," 141-55.

19. Marsden et al., *Constructing the Countryside*, 75.

20. For works on the bird protection movement in Britain, see Doughty, *Feather Fashions and Bird Preservation*; Allen, *The Naturalist in Britain*, 195-206; Gates, *Kindred Nature*, 113-43; Bonhomme, "Nested Interests," 47-68.

21. Bonhomme, "Nested Interests," 62.

22. See Lambert and Lester, *Colonial Lives across the British Empire*. This approach is similar to recent works on translocality, which adopt Doreen Massey's notion of place as relational and mutually affecting. For works on translocal natures, see Raffles, *In Amazonia*; Cameron and Matless, "Translocal Ecologies," 1-27.

23. See especially the chapter titled "Home, Nation, and Empire," in Blunt and Dowling, *Home*, 114, 159.

24. Cole, *The Story of Aldershot*, 25-28, 37.

25. Murray, *A Handbook for Travellers*, 174-75.

26. Tivers, "'The Home of the British Army,'" 303-19. Rachel Woodward studies these ideas in a contemporary context. See Woodward, *Military Geographies*.

27. Richardson, *The Anglo-Indian Passage*, viii.

28. Richardson, ix.

29. The Hundredth Regiment was stationed at Gibraltar from 1859 until 1863, whence it moved to Malta. Boulton, *Reminiscences of the North-West Rebellions*, 18-19.

30. A mid-nineteenth-century military album located in Ottawa likely belonged to a soldier-officer in the Rifle Brigade. Images include scenes from Canada, the Crimea, Ireland, and Aldershot, England. Rifle Brigade Album, 1948-027 PIC, Library and Archives Canada, Ottawa, Canada.

31. Arnold, "Report on the Climate of Aldershot Camp," 15.

32. Driver and Martins, *Tropical Visions in an Age of Empire*, 3.

33. Cole, *The Story of Aldershot*, 30-32.

34. Cole, 55-63.

35. Homans, "'To the Queen's Private Apartments,'" 1-41.

36. Homans, 46-52.

37. Boulton, *Reminiscences of the North-West Rebellions*, 18-19.

38. See also Sharp, "Gendering Nationhood," 97-108.

39. Philip Savile Grey Reid was also known as Savile Reid in his published works. Previous to Woolwich, he was at the Burney's Naval School in Gosport, Hampshire County. Warr, *Manuscripts and Drawings*. Burney's Naval School was a popular site for

young boys interested in joining the Royal Navy. Charles Perry, son of Sir W. Edward Perry, attended Burney's Naval School in the 1840s when he was twelve. His father was a Royal Navy officer and superintendent at the Haslar Hospital in Gosport. See Parry, *Memorials of Charles Parry*, 1-3.

40. Blunt, "Imperial Geographies of Home," 423.

41. Philip Savile Grey Reid, 22 April 1877, *MSS Stray Notes on Ornithology, 1871-1890*, Philip Savile Grey Manuscript Collection, Natural History Museum Ornithological Library, Tring, UK.

42. Reid, *MSS Stray Notes on Ornithology*, 22 April 1877.

43. Buettner, *Empire Families*, 214.

44. Sclater, *List of the Vertebrated Animals*, 287, 286.

45. By the early 1880s, Reid was among one of many British military officers who were elected to the BOU. For a list of a sample of individuals and their election years, see *Ibis* 1 (1883): 1874, H. H. Godwin-Austen, lieutenant colonel, FES, FZS; 1870, Leonard Howard Lloyd Irby, lieutenant colonel, FZS; 1880, Henry Robert Kelham, lieutenant, Seventy-Fourth Highlanders; 1874, Alexander W. M. Clark Kennedy, FLS, FRGS,; 1875, Paget Walter Le Strange, lieutenant colonel, Royal Artillery; 1874, John Hayes Lloyd, major, FZS; 1873, Sir Oliver Beauchamp Coventry St. John, colonel, Royal Engineers; 1870, G. Ernest Shelley, FZS, late captain, Grenadier Guards.

46. Reid, *MSS Stray Notes on Ornithology*, 22 May 1877.

47. Irby, *British Birds: Key List*, i. Many of the founding members were active in the maintenance of empire, such as the first president, Henry Maurice Drummond Hay, who served with the Forty-Second Regiment (Black Watch) in Bermuda, Nova Scotia, the Ionian Islands, and Malta. Other members included Edward Newton (1832-97), a brother of Professor Alfred Newton and a colonial administrator, serving as colonial secretary in Mauritius (1859-77) and as colonial secretary and lieutenant governor of Jamaica. The Newton family also owned plantations in the West Indies.

48. Reid, *MSS Stray Notes on Ornithology*. BOU meetings were located at "the den" at 6 Tenterden Street, Hanover Square, which was then the headquarters of the BOU but also home to Lord Lilford, a member in London. The Tenterden Street site was shared by other members of the BOU, including Salvin and Godman, who used the house for their "museum and library" of Latin American birds. Godman and Salvin, *Biologia Centrali-Americana*, 7.

49. Reid, *MSS Stray Notes on Ornithology*, 30 November 1877.

50. Reid, *MSS Stray Notes on Ornithology*, 30 November 1877. Irby and Reid served together at Gibraltar in 1871-73. Anon., "Obituary: Philip Savile Grey Reid," 365-67.

51. Reid, *MSS Stray Notes on Ornithology*, 28 October 1876.

52. Reid, *MSS Stray Notes on Ornithology*, 15 August 1876. See Kelham, "Ornithological Notes," 362-95.

53. Feilden joined the Narnes Arctic expedition and would later serve with Reid at Natal in 1881. Reid, *MSS Stray Notes on Ornithology*, 18 June 1877.

54. Reid, *MSS Stray Notes on Ornithology*, 31 March 1878.

55. Reid, *MSS Stray Notes on Ornithology*, 31 March 1878.

56. Southampton served as a "gateway to empire," with the traffic of ships to and from different parts of the empire; it was also a site in the making of an imperial

scientific hero, David Livingstone, who returned home from Africa in 1856, and again in 1874, when his body traveled back to London for his funeral. Taylor, *Southampton*; Lewis, "Southampton and the Making of an Imperial Myth," 31–46; Pettitt, *Dr. Livingstone, I Presume?*, 31, 124.

57. Reid, *MSS Stray Notes on Ornithology*, 11 October 1877.

58. Marquis of Ruvigny and Raineval, *Plantagenet Roll of the Blood Royal*, 353; Reid, *MSS Stray Notes on Ornithology*, 30 August 1878.

59. Reid, *MSS Stray Notes on Ornithology*, 28 December 1877.

60. Matless, "Moral Geographies of English Landscape," 141–55.

61. Tebbutt, "Rambling and Manly Identity," 1126. See Lowe, Murdoch, and Cox, "A Civilised Retreat?," 63–83.

62. Kingsley, *The Study of Natural History*, 3.

63. Kingsley, 9.

64. Kingsley, 3.

65. David Matless provides examples of the various "bodies of England" in Matless, *Landscape and Englishness*, 87. M. Tebbutt discusses changes in English masculinities in the countryside in the late 1880s. See Tebbutt, "Rambling and Manly Identity," 1125–53.

66. Reid, *MSS Stray Notes on Ornithology*, 26 March 1878.

67. Reid, *MSS Stray Notes on Ornithology*, 26 March 1878.

68. Many of the original trees planted in the 1860s are still at Aldershot. Reid, *MSS Stray Notes on Ornithology*, 26 March 1878.

69. On 8 October, 1876, Reid stated in his journal how he and Denison walked to Fleet Pond: "Our walk proved quite an ornithological success, for we saw five *Wild Ducks*, (being Sunday we had no gun) and two or three *Cial Buntings* near the pond." See also Reid, *MSS Stray Notes on Ornithology*, 23 August 1877 and 27 September 1877.

70. Reid, *MSS Stray Notes on Ornithology*, 27 September 1877.

71. For a discussion of anthropogenic natures, see Cameron, "Distinguished Historical Geography Lecture 2010," 5–22.

72. Murray, *A Handbook for Travellers*, 172.

73. Between 1830 and 1900, more than a hundred editions were published, making White's book a familiar item on Victorian bookshelves. See Lipscomb, "Introducing Gilbert White," 552.

74. White, *The Natural History of Selborne*, 109–12.

75. Reid, *MSS Stray Notes on Ornithology*, 6 May 1877.

76. Webster et al., "Links between Worlds," 77. Hugh Raffles uses Latour's analogy of the network of railway tracks in explaining this sense of place. See Raffles, *In Amazonia*, 181–82.

77. Bewick, *A History of British Birds*, viii; Crang, "Placing Jane Austen, Displacing England," 111–32.

78. Reid, *MSS Stray Notes on Ornithology*, 9 April 1876.

79. Reid, *MSS Stray Notes on Ornithology*, 9 April 1876.

80. Reid, *MSS Stray Notes on Ornithology*, 17 February 1878.

81. See page 2 of Reid's field journal entry for 1871. Reid, *MSS Stray Notes on Ornithology*.

82. Reid, *MSS Stray Notes on Ornithology*, 25 December 1873.

83. Professor Alfred Newton presented a paper on the protection of wild birds of prey and seafowl at the British Association for the Advancement of Science (BAAS) meeting at Norwich in 1868, in which he condemned the "wholesale slaughter of many of our birds during the breeding season" that would "result in their extinction, unless laws were passed to give them protection." See Wollaston, *Life of Alfred Newton*, 139–40.

84. MacDonald, "What Makes You a Scientist," 55.

85. Reid returned to England in 1875 after serving in Bermuda. Reid, *MSS Stray Notes on Ornithology*, 24 June 1875.

86. MacDonald, "What Makes You a Scientist," 58.

87. Sharpe, *Hand-Book to the Birds of Great Britain*, v.

88. Allen provides a historical context on birds and census work in *The Naturalist in Britain*.

89. Newton, "On the Possibility of Taking an Ornithological Census," 191.

90. Newton, 192. For more information on the British zoological census, see Allen, *The Naturalist in Britain*, 218. For details on the census of the British Empire, see Christopher, "The Quest for a Census of the British Empire," 268–85.

91. Born in London, More lived on the Isle of Wight until 1867 but made frequent visits to Ireland to study the flora. He published *Cybele Hibernica* (1866). In his zoological work, More concentrated mainly on birds. He was assistant curator at the Dublin Natural History Museum, 1867–81, and curator, 1881–87. More, "On the Distribution of Birds of Great Britain," 1.

92. More, 4.

93. Walters, *A Concise History of Ornithology*, 151.

94. Gould, *An Introduction to the Birds of Great Britain*, 3.

95. Gould, 3.

96. Gould, 8.

97. Camerini, "Evolution, Biogeography, and Maps," 705–7.

98. Adams and Adams, *On Ornithology as a Branch of Liberal Education*, 33.

99. Adams, *Notes of a Naturalist*, 19.

100. See page 22 of Reid's field journal entry for 1871. Reid, *MSS Stray Notes on Ornithology*.

101. Yarrell, Newton, and Saunders, *A History of British Birds*, vii. Alfred Newton and Howard Saunders revised Yarrell's classic work in 1874.

102. Yarrell, Newton, and Saunders, vi.

103. Yarrell, Newton, and Saunders, vi.

104. British Ornithologists Union, *A List of British Birds*, v. For more than a hundred years the BOU has maintained a list of birds that have been recorded in Britain and Ireland.

105. According to the BOU list, "residents" were those that could be "found throughout the year, and actually breed in some part of the British Islands"; "summer visitors" were species that visited the islands in the summer and bred "within these confines"; "winter visitors" were those that visited Britain in winter but did not breed there; and "occasional visitors" were those that occurred irregularly but were encountered in the British Isles "with more or less frequency." See British Ornithologists Union, *A List of British Birds*.

106. Kelsall, "A Briefly Annotated List of the Birds of Hampshire," 90.

107. Allen, *The Naturalist in Britain*, 218.

108. Anon., "Obituary: Philip Savile Grey Reid," 366.

109. Reid, "The Birds of the Bermudas," 8, 13, 18.

110. Kelsall, "A Briefly Annotated List of the Birds of Hampshire," 91. Kelsall lived at Farnham, a town close to Aldershot.

111. Kelsall, 90. Captain Henry Hadfield published several articles in the *Zoologist* on the birds of Kingston, Canada West, when stationed at Fort Henry.

112. Kelsall, 114.

113. Kelsall, 113–15. Lieutenant Colonel Henry Wemyss Feilden (1838–1921) also collected birds at Aldershot in 1878, including a specimen now housed at the Manchester Museum.

114. George Montagu served in British North America and published *Ornithological Dictionary; or Alphabetical Synopsis of British Birds* (1802) and *Supplement* (1813); he was one of the earliest members of the Linnaean Society. Mearns and Mearns refer to these works in *The Bird Collectors*, 188–89; M. G. W., "Montagu, George (1751–1815)," 693–94; Mearns and Mearns, *Biographies for Birdwatchers*, 263–70.

115. Yarrell, Newton, and Saunders, *A History of British Birds*, 139–40.

116. Yarrell, Newton, and Saunders, 141.

117. Irby, *Ornithology of the Straits of Gibraltar*, 167.

118. Irby, 167.

119. Reid, *MSS Stray Notes on Ornithology*, 17 February 1878.

120. Reid, *MSS Stray Notes on Ornithology*, 30 September 1877.

121. Yarrell, Newton, and Saunders, *A History of British Birds*, 399.

122. Yarrell, Newton, and Saunders, 399.

123. Reid, *MSS Stray Notes on Ornithology*, 31 May 1878.

124. Reid, *MSS Stray Notes on Ornithology*, 31 May 1878.

125. British Museum (Natural History), *Guide to the Gallery of Birds*. For information on the nest collection at the Natural History Museum, see Russell, Hansell, and Reilly, "Constructive Behaviour." The series was painted in 1888 by Emily Mary Bibbens Warren (1869–1956) and is housed at the archives at the Natural History Museum, London, UK (NHM 1166/05).

126. In the preface of British Museum (Natural History), *Guide to the Gallery of Birds*.

127. British Museum (Natural History), *Guide to the Gallery of Birds*, 18–19.

128. British Museum (Natural History), *Guide to the Gallery of Birds*, 14.

129. S. D. W. "Art. II. *On the Habits, Haunts, and Nidification of the Robin Redbreast*," 6.

130. Sharpe, "Ornithology at South Kensington," 170.

131. Reid, *MSS Stray Notes on Ornithology*, 16 April 1876.

132. Anon., "Obituary: Philip Savile Grey Reid," 365–67.

## Chapter Six

1. William Notman thus described the life of Lieutenant Colonel William Rhodes, a British military officer who settled in Quebec and maintained an interest in birds. Notman, *Portraits of British Americans*, 74.

2. Star and Griesemer, "Institutional Ecology, 'Translations' and Boundary Objects," 387–420.

3. Jacobs, "The Intimate Politics of Ornithology in Colonial Africa," 565.

4. For an extended version on this literature, see Greer, "Geopolitics and the Avian Imperial Archive."

5. Driver, "Sub-merged Identities," 412.

6. Linda Colley discusses how "Britishness" was a Scottish invention. See Colley, *Britons*, 120–32.

7. Adams, *Field and Forest Rambles*, 12.

## Afterword

1. Sears, "Michael Palin."

2. Sears.

3. For an overview of the bird hunting issue in Malta, see Fenech, *Fatal Flight*. See also Greer, "Transnational Ecologies in Malta."

4. McCarthy, *Say Goodbye to the Cuckoo*, 19–20.

5. McCarthy, 19–20.

6. European Commission, "The Birds Directive"; European Commission, "Environment."

7. TVM, "Birdlife Malta Loses Libel Case."

8. I gained much insight on these issues from Charles Farrugia, national archivist of Malta.

9. Baldacchino, "A Nationless State?," 195.

10. Mitchell, *Ambivalent Europeans*, 9.

11. I was fortunate to meet John Borg, curator of the National Museum of Natural History Malta, and Charles Farrugia, both of whom helped me in my researches.

12. Chadha, "Ambivalent Heritage," 339.

13. BirdLife International. "BirdLife Malta Raptor Camp."

14. Agnew, *Reinventing Geopolitics*, 47–48.

15. Bhabha, "The Other Question," 18.

16. Gregory, *The Colonial Present*, 10.

# *Bibliography*

Archival Collections

Cambridge, United Kingdom
    Cambridge University Library
        Alfred Newton Papers
        Royal Commonwealth Society Collections
London, United Kingdom
    Library and Archives of the Royal Society
        Royal Society Correspondence
        Thomas Davies Papers
    Library of the Zoological Society of London
        John Drummond Hay Papers
        L. Howard Irby Papers
        Zoological Society of London Correspondence
    National Army Museum
        Cundall and Howlett Series
    Natural History Museum Library and Archives
        Emily Mary Bibbens Warren Watercolour Series
        John Gould Correspondence
    The Royal Collection
        Cundall and Howlett Series
    The Royal Institution of Great Britain
        John Davy Collection
    Victoria and Albert Museum
New Haven, CT
    Yale Center for British Art
        Paul Mellon Collection
Norwich, United Kingdom
    Reid Private Collection (Dave Reid)
        Philip Savile Grey Reid Papers
Ottawa, Canada
    Library and Archives Canada
        Arthur Dresser Fonds
        Rifle Brigade Album
Tring, United Kingdom
    Natural History Museum Ornithological Library
        Charles Augustus Wright Manuscript Collection
        Philip Savile Grey Manuscript Collection

Other Sites Consulted

The British Library, London, UK
Firepower, Royal Artillery Museum, Woolwich, UK
Gibraltar Garrison Library
Halifax Citadel National History Site, Halifax, Canada
Malta Garrison Library
Marischal Museum, Aberdeen, UK
National Army Museum, London, UK
National Museum of Natural History, Mdina, Malta
The Royal Geographical Society, London, UK
The Royal Highland Fusiliers Regimental Museum, Glasgow, UK
St. Barbe Museum and Art Gallery, Lymington, UK

Dictionaries and Encyclopedias

Baynes, T. S. "Malta." In *Encyclopaedia Britannica* 15 (1888): 340.
Buggey, Susan. "Downs, Andrew." In *Dictionary of Canadian Biography 1891-1900*. Vol. 12. 268-69. University of Toronto/Université Laval, 2003. http://www.biographi.ca/en/bio/downs_andrew_12E.html. Accessed 17 July 2019.
Casey, Ellen M. "Adams, Bertha Jane Leith (1837-1912)." In *Oxford Dictionary of National Biography*. Oxford University Press, 2004-. https://doi.org/10.1093/ref:odnb/55792. Accessed 19 July 2011.
Chichester, H. M. "Napier, Edward Delaval Hungerford Elers (1808-1870)," revised by James Falkner. In *Oxford Dictionary of National Biography*. Oxford University Press, 2004-. https://doi.org/10.1093/ref:odnb/19750. Accessed 19 July 2011.
Edwards, J. C. "Sclater, Philip Lutley (1829-1913)." In *Oxford Dictionary of National Biography*. Oxford University Press, 2004-. https://doi.org/10.1093/ref:odnb/38295. Accessed 19 July 2011.
Gaston, Anthony J. "Adams, Andrew Leith (1827-1882)." In *Oxford Dictionary of National Biography, 1861-1870*. Oxford University Press, 2004-. https://doi.org/10.1093/ref:odnb/111. Accessed 19 July 2011.
Johnson, R. E. "Richardson, Sir John." In *Dictionary of Canadian Biography*. Vol. 9. University of Toronto/Université Laval, 2003-. http://www.biographi.ca/en/bio/richardson_john_9E.html. Accessed 17 July 2019.
Johnson, Robert E., and Margaret H. Johnson. "Richardson, Sir John (1787-1865)." In *Oxford Dictionary of National Biography*. Oxford University Press, 2004-. https://doi.org/10.1093/ref:odnb/23568. Accessed 28 January 2011.
Keene, H. G. "Hay, Arthur, Ninth Marquess of Tweeddale (1824-1878)," revised by Yolanda Foote. In *Oxford Dictionary of National Biography*. Oxford University Press, 2004-. https://doi.org/10.1093/ref:odnb/12710. Accessed 28 January 2011.
McAlpine, Donald F. "Boardman, George Augustus." In *Dictionary of Canadian Biography 1901-1910*. Vol. 13. University of Toronto/Université Laval, 2003-. http://www.biographi.ca/en/bio/boardman_george_augustus_13E.html. Accessed 17 July 2019.

Roberts, A. D. "Livingstone, David (1813-1873)." In *Oxford Dictionary of National Biography* Oxford University Press, 2004-. https://doi.org/10.1093/ref:odnb /16803. Accessed 28 January 2011.

Tasker, Meg. "Adams, Francis William Lauderdale (1862-1893)." In *Oxford Dictionary of National Biography*. Oxford University Press, 2004-. https://doi.org/10.1093 /ref:odnb/114. Accessed 19 July 2011.

Taylor, Hugh A. "Harding, Francis Pym." In *Dictionary of Canadian Biography, 1871-1880*. Vol. 10. University of Toronto/Université Laval, 2003-. http://www .biographi.ca/en/bio/harding_francis_pym_10E.html. Accessed 17 July 2019.

Vetch, R. H. "Portlock, Joseph Ellison (1794-1864)," revised by Elizabeth Baigent. In *Oxford Dictionary of National Biography*. Oxford University Press, 2004-. https://doi .org/10.1093/ref:odnb/22585. Accessed 19 July 2011.

Dissertations

Kohout, Amy. "From the Field: Nature and Work on American Frontiers, 1876-1909." PhD diss., Cornell University, 2015.

Kunstmann, John G. "The Hoopoe: A Study in European Folklore." PhD diss., University of Chicago, 1938.

Presentations

Greer, K. "Birds and Biography. Writing the 'Life Geography' of Military-Surgeon Andrew Leith Adams (1827-1882), 22nd Regiment of Foot." Paper presented at the annual meeting of the Canadian Historical Association, Montreal, 30 May to 1 June, 2010.

———. "The Colonial Ornithology of Thomas Wright Blakiston (1832-1891)." Paper presented at the HGRG Session: Indigenous Knowledge, Resistance and Agency: Telling the Hidden Histories of Geographical Field Science and Exploration, annual conference of the Royal Geographical Society with the Institute of British Geographers (RGS-IBG), London, 27-29 August 2008.

———. "Ornithology on 'Old Gib.'" Paper presented at the CGSG and HGSG Session: Memories, Memorials and Identities at the annual general meeting of the American Association of Geographers, Boston, 15-19 April 2008.

———. "Ornithology on 'the Rock.'" Paper presented at the McGill-Queen's Graduate Student Conference in History, Kingston, Ontario, 13-14 March 2009.

Books

Adams, Andrew L. *Field and Forest Rambles, with Notes and Observations on the Natural History of Eastern Canada*. London: H. S. King & Co., 1873.

———. *Notes of a Naturalist in the Nile Valley and Malta*. Edinburgh: Edmonston & Douglas, 1870.

———. *Wanderings of a Naturalist in India: The Western Himalayas, and Cashmere*. Edinburgh: Edmonston & Douglas, 1867.

Adams, Francis. *The Genuine Works of Hippocrates*. New York: William Wood, 1891.
Adams, Francis, and Andrew L. Adams. *On Ornithology as a Branch of Liberal Education: Containing Notes of All the Wild Birds Which Have Been Discovered in Banchory Ternan, with Remarks on Such of Them as Have Been Found in India*. Aberdeen, UK: John Smith, 1859.
Agnew, John. *Reinventing Geopolitics: Geographies of Modern Statehood* (Hettner-Lectures). Stuttgart: Franz Steiner Verlag, 2001.
Ainsworth, William. *Researches in Assyria, Babylonia, and Chaldaea; Forming Part of the Labours of the Euphrates Expedition*. London: J. W. Parker, 1838.
Alberti, Samuel J. M. M. *Nature and Culture: Objects, Disciplines and the Manchester Museum*. Manchester: University of Manchester Press, 2009.
Alexander, James E. *Passages in the Life of a Soldier, or, Military Service in the East and West*. London: Hurst & Blacket, 1857.
Allen, D. W. "Young England: Muscular Christianity and the Politics of the Body in Tom Brown's Schooldays." In *Muscular Christianity: Embodying the Victorian Age*, edited by Donald E. Hall, 114–32. Cambridge: Cambridge University Press, 1994.
Allen, David E. *The Naturalist in Britain: A Social History*. London: A. Lane, 1976.
Allen, Elsa G. *The History of American Ornithology before Audubon*. New York: Russell & Russell, 1951.
Anderson, Benedict. "Census, Map, Museum." In *Imagined Communities: Reflections on the Origin and Spread of Nationalism*, 163–85. London: Verso, 1991.
Angell, Norman. *The Defence of the Empire*. New York: D. Appleton-Century, 1938.
Anon. *Gibraltar and Its Sieges*. London: Thomas Nelson & Sons, 1879.
———. *Penny Cyclopaedia of the Society for the Diffusion of Useful Knowledge*. London: Charles Knight, 1843.
Archer, Edward G. *Gibraltar, Identity and Empire*. London: Routledge, 2006.
Army Medical Department. *A Catalogue of the Collection of Mammalia and Birds in the Museum of the Army Medical Department at Fort Pitt, Chatham*. Chatham, UK: Army Medical Department, 1838.
Arnold, David. *Colonizing the Body: State Medicine and Epidemic Disease in Nineteenth-Century India*. Berkeley: University of California Press, 1993.
———. *The Tropics and the Traveling Gaze: India, Landscape, and Science 1800–1856*. Seattle: University of Washington Press, 2006.
Bartholomew, J. G. *Bartholomew's Physical Atlas*. Vol. 5, *Atlas of Zoogeography*. Edinburgh: John Bartholomew, 1911.
Bartlett, William H. *Gleanings on the Overland Route: Pictorial and Antiquarian*. London: Hall, Virtue & Co., 1851.
Batten, L. A., C. J. Bibby, and P. Clement. *Red Data Birds in Britain*. London: T & AD Poyser, 2013.
Bayly, Christopher A. *Empire and Information: Intelligence Gathering and Social Communication in India, 1780–1870*. Cambridge: Cambridge University Press, 1996.
Beattie, James, Edward Melillo, and Emily O'Gorman. "Eco-cultural Networks in the British Empire, 1860–1940." In *Eco-cultural Networks and the British Empire: New Views on Environmental History*, edited by James Beattie, Edward Melillo, and Emily O'Gorman, 3–20. London: Bloomsbury, 2014.

Becker, Cynthia J. *Amazigh Arts in Morocco: Women Shaping Berber Identity*. Austin: University of Texas Press, 2006.

Berger, Carl. *Science, God and Nature in Victorian Canada*. Toronto: University of Toronto Press, 1983.

Bewell, Alan. "Jefferson's Thermometer: Colonial Biogeographical Constructions of the Climate." In *Romantic Science: The Literary Forms of Natural History*, edited by Noah Heringman, 111–38. Albany: State University of New York Press, 2003.

———. *Romanticism and Colonial Disease*. Baltimore: Johns Hopkins University Press, 1999.

Bewick, Thomas. *A History of British Birds: Containing the History and Description of Water Birds*. Newcastle, UK: Longman & Co., 1826.

Bircham, Peter. *A History of Ornithology*. London: HarperCollins, 2007.

Bird, Will R. *North Shore New Brunswick Regiment*. Fredericton, NB: Brunswick Press, 1963.

Blackwood, Alicia. *A Narrative of Personal Experiences and Impressions during My Sojourn in the East throughout the Crimean War*. London: Hatchard, 1881.

Blakiston, John. *Twelve Years' Military Adventure in Three Quarters of the Globe*. Vol. 1. London: Henry Colburn, 1829.

———. *Twenty Years in Retirement*. London: James Cochrane, 1835.

Blunt, Alison. *Travel, Gender, and Imperialism: Mary Kingsley and West Africa*. New York: Guilford Press, 1994.

Blunt, Alison, and Robyn Dowling. *Home*. London: Routledge, 2006.

Borzesi, Giuseppe P. *The Historical Guide to the Island of Malta and Its Dependencies*. Malta: Government Press, 1830.

Boulton, Charles A. *Reminiscences of the North-West Rebellions*. Toronto: Grip Printing & Pub Co., 1886.

Bourguet, M.-N., B. Lepetit, D. Nordman, and M. Sinarellis, eds. *L'invention scientifique de la Méditerranée*. Paris: École des hautes études en sciences sociales, 1998.

Boyd, Hilton. "Manliness and the Mid-Victorian Temperament." In *The Blind Victorian Henry Fawcett and British Liberalism*, edited by Lawrence Goldman, 60–70. Cambridge: Cambridge University Press, 1989.

Braudel, Ferdinand. *The Mediterranean and the Mediterranean World in the Age of Philip II*. Vol. 1. London: HarperCollins, 1972.

Bree, Charles R. *A History of the Birds of Europe, Not Observed in the British Isles*. Vol. 1. London: Groombridge, 1859.

British Museum. *The History of the Collections Contained in the Natural History Departments of the British Museum*. Vol. 2. London: British Museum, 1906.

British Museum of Natural History. *Guide to the Gallery of Birds in the Department of Zoology, British Museum (Natural History) Part 2: Nesting-Series of British Birds*. London: Trustees of the British Museum, 1919.

British Ornithologists Union. *A List of British Birds*. London: J. van Voorst, 1883.

Brown, Robert. *Our Earth and Its Story: A Popular Treatise on Physical Geography*. London: Cassell, 1889.

Browne, Janet. *The Secular Ark: Studies in the History of Biogeography*. New Haven, CT: Yale University Press, 1983.

Buettner, Elizabeth. *Empire Families: Britons and Late Imperial India*. Oxford: Oxford University Press, 2004.

Burant, Jim. *Drawing on the Land: The New World Travel Diaries and Watercolours of Milicent Mary Chaplin, 1838-1842*. Newcastle, ON: Penumbra Press, 2004.

———. *Friendly Spies on the Northern Tour, 1815-1837*. Ottawa: Ministry of Supply and Services, Canada, 1981.

Burroughs, Peter. "An Unreformed Army? 1815-1868." In *The Oxford Illustrated History of the British Army*, edited by David Chandler and Ian Beckett, 160-88. Oxford: Oxford University Press, 1994.

Burton, Lesley, and Beryl Peacey. *Soldiers of the Queen: Gosport as a Garrison Town*. Gosport, UK: Gosport Historical Society, 2002.

Campbell, George. *The British Empire*. London: Cassell, 1887.

Caniage, Jean. "France, England, and the Tunisian Affair." In *France and Britain in Africa: Imperial Rivalry and Colonial Rule*, edited by Prosser Gifford and William Roger Louis, 35-72. New Haven, CT: Yale University Press, 1971.

Cannadine, David. *Ornamentalism: How the British Saw Their Empire*. New York: Oxford University Press, 2001.

Chesney, Francis R. *Narrative of the Euphrates Expedition: Carried on by Order of the British Government during the Years 1835, 1835, and 1837*. London: Longmans, Green, & Co., 1868.

Chirovsky, Nicholas L. *An Introduction to Russian History*. Oxford: Oxford University Press, 1967.

Cohn, Bernard S. *Colonialism and Its Forms of Knowledge: The British in India*. Princeton, NJ: Princeton University Press, 1996.

Cole, Howard N. *The Story of Aldershot: A History and Guide to Town and Camp*. Aldershot, UK: Gale & Polden, 1951.

Colley, Linda. *Britons: Forging the Nation 1707-1837*. New Haven, CT: Yale University Press, 1992.

———. *Captives: Britain, Empire and the World, 1600-1850*. London: Pimlico, 2002.

Collingham, E. M. *Imperial Bodies: The Physical Experience of the Raj, c. 1800-1947*. Cambridge: Polity Press, 2001.

Conacher, J. B. *Britain and the Crimea, 1855-56: Problems of War and Peace*. London: Macmillan, 1987.

Crang, Mike. "Placing Jane Austen, Displacing England: Touring between Book, History and Nation." In *Jane Austen and Co.: Remaking the Past in Contemporary Culture*, edited by S. Pucci and J. Thompson, 111-32. New York: State University of New York Press, 2003.

Crispo-Barbaro, G. *The Birds of Malta: With Their Maltese, Systematic, Italian and English*. Malta: Malta News, 1878.

Curtin, Philip D. *Death by Migration: Europe's Encounter with the Tropical World in the Nineteenth Century*. Cambridge: Cambridge University Press, 1989.

———. "Disease and Imperialism." In *Warm Climates and Western Medicine: The Emergence of Tropical Medicine, 1500-1900*, edited by David Arnold, 99-107. Amsterdam: Rodopi, 1996.

Cushman, Gregory. *Guano and the Opening of the Pacific World: A Global Ecological History*. Cambridge: Cambridge University Press, 2013.

Dallison, Robert L. *Turning Back the Fenians: New Brunswick's Last Colonial Campaign*. Fredericton, NB: Goose Lane Editions, 2006.

Davis, Diana K. *Resurrecting the Granary of Rome: Environmental History and French Colonial Expansion in North Africa*. Athens: Ohio University Press, 2007.

Dawson, Graham. *Soldier Heroes: British Adventure, Empire, and the Imagining of Masculinities*. London: Routledge, 1994.

Delborough, James. *Collecting the World: Hans Sloane and the Origins of the British Museum*. Cambridge, MA: Belknap Press of Harvard University Press, 2017.

Dickenson, Victoria. *Drawn from Life: Science and Art in the Portrayal of the New World*. Toronto: Toronto University Press, 1998.

Doughty, Robin. *The English Sparrow in the American Landscape: A Paradox in Nineteenth-Century Wildlife Conservation*. Oxford: University of Oxford Press, 1978.

———. *Feather Fashions and Bird Preservation: A Study in Nature Protection*. Berkeley: University of California Press, 1975.

Drinkwater, John. *A History of the Siege of Gibraltar*. London: T. Spilsbury, 1786.

Driver, Felix. *Geography Militant: Cultures of Exploration and Empire*. Oxford: Blackwell, 2001.

Driver, Felix, and Luciana Martins, eds. *Tropical Visions in an Age of Empire*. Chicago: University of Chicago Press, 2005.

Edney, Matthew H. *Mapping an Empire: The Geographical Construction of British India, 1765–1843*. Chicago: University of Chicago Press, 1997.

Edwards, Elizabeth, Chris Gosden, and Ruth B. Phillip, eds. *Sensible Objects: Colonialism, Museums, and Material Culture*. Oxford: Berg, 2006.

Elder, G., J. Wolch, and J. Emel. "Le Practique Sauvage: Race, Place, and the Human-Animal Divide." In *Animal Geographies: Place, Politics, and Identity in the Nature-Culture Borderlands*, edited by J. Wolch and J. Emel, 241–50. London: Verso, 1998.

Evans, Emyr E. *Ireland and the Atlantic Heritage: Selected Writings*. Dublin: Lilliput Press, 1996.

Fan, Fa-Ti. *British Naturalists in Qing China: Science, Empire, and Cultural Encounter*. Cambridge, MA: Harvard University Press, 2004.

Farnie, D. A. *East and West of Suez: The Suez Canal in History, 1854–1956*. Oxford: Clarendon Press, 1969.

Fenech, Natalino. *Fatal Flight: The Maltese Obsession with Killing Birds*. Shrewsbury, UK: Quiller Press, 1992.

Ferguson-Lees, James, and David A. Christie. *Raptors of the World*. London: Christopher Helm, 2001.

Finlayson, Clive. *Birds of the Strait of Gibraltar*. London: T & AD Poyser, 1992.

———, ed. *Gibraltar: 300 Years of Images*. Gibraltar: Government of Gibraltar, 2004.

Finnegan, Diarmid A. *Natural History Societies and Civic Culture in Victorian Scotland*. London: Pickering & Chatto, 2009.

Fisher, Clemency, ed. *A Passion for Natural History: The Life and Legacy of the 13th Earl of Derby*. Liverpool: National Museums and Galleries on Merseyside, 2002.

Forsyth, Isla. "Biography and the Military Archive." In *The Routledge Companion to Military Research Methods Routledge Handbooks*, edited by Alison J. Williams, Neil Jenkings, Rachel Woodward, and Matthew F. Rech, 44–57. London: Routledge, 2016.

Foucault, Michel. *Discipline and Punish: The Birth of the Prison*. 2nd ed. New York: Vintage Books, 1979.

———. *The Order of Things: An Archaeology of the Human Sciences*. New York: Pantheon Books, 1970.

Franzen, Jonathan. "Emptying the Skies." *New Yorker*, July 26, 2010.

Freller, Thomas. *Malta and the Grand Tour*. Santa Venera, Malta: Midsea Books, 2009.

French, David. *Military Identities: The Regimental System, the British Army, and the British People c. 1870–2000*. Oxford: Oxford University Press, 2005.

Gambi, Lucio. "Geography and Imperialism in Italy: From the Unity of the Nation to the 'New' Roman Empire." In *Geography and Empire*, edited by Neil Smith and Anne Marie Godlewska, 74–91. Oxford: Blackwell, 1994.

Garratt, Geoffrey T. *Gibraltar and the Mediterranean*. New York: Coward McCann, 1939.

Gates, Barbara. *Kindred Nature: Victorian and Edwardian Women Embrace the Living World*. Chicago: University of Chicago Press, 1998.

Gibraltar Garrison Library. *Catalogue of the Books in the Gibraltar Garrison Library*. Gibraltar: Gibraltar Chronicle, 1876.

Gilbard, George J. *A Popular History of Gibraltar, Its Institutions, and Its Neighbourhood on Both Sides of the Straits*. Gibraltar: Gibraltar Garrison Library, 1881.

Godlewska, Anne M. C. "Humboldt's Visual Thinking: From Enlightenment Vision to Modern Science." In *Geography and Enlightenment*, edited by David Livingstone and Charles Withers, 236–79. Chicago: University of Chicago Press, 1999.

Godman, Frederick DuCane, and Osbert Salvin. *Biologia Centrali-Americana; Or, Contributions to the Fauna and Flora of Mexico and Central America: Introductory Volume*. Privately published, 1915.

Goldfrank, David M. *The Origins of the Crimean War*. London: Longman, 1994.

Gordon, Arthur. "Wilderness Journeys in New Brunswick." In *Vacation Tourists and Notes of Travel in 1860 [1861], [1862-3]*, edited by Francis Galton, 457–524. London: MacMillan, 1864.

Gould, John. *An Introduction to the Birds of Great Britain*. London: Taylor & Francis, 1873.

Green, Martin B. *Dreams of Adventure, Deeds of Empire*. New York: Basic Books, 1979.

Gregory, Derek. *The Colonial Present: Afghanistan, Palestine, Iraq*. Oxford: Blackwell, 2004.

———. "Emperors of the Gaze: Photographic Practices and Productions of Space in Egypt, 1839–1914." In *Picturing Place: Photography and the Geographical Imagination*, edited by Joan M. Schwartz and James R. Ryan, 195–225. London: I. B. Tauris, 2003.

Gupta, Akhil, and James Ferguson. "Discipline and Practice: 'The Field' as Site, Method, and Location in Anthropology." In *Anthropological Locations: Boundaries and Grounds of a Field Science*, edited by Akhil Gupta and James Ferguson, 1–46. Berkeley: University of California Press, 1997.

Haley, Bruce. *The Healthy Body and Victorian Culture*. Cambridge, MA: Harvard University Press, 1978.

Hannah, Matt. *Governmentality and the Mastery of Territory in Nineteenth-Century America*. Cambridge: Cambridge University Press, 2000.

Haraway, Donna J. *Primate Visions: Gender, Race, and Nature in the Modern World of Science*. New York: Routledge, 1989.

Hargreaves, John D. *Aberdeenshire to Africa: Northeast Scots and British Overseas Expansion*. Aberdeen, UK: Aberdeen University Press, 1981.

Harrison, Mark. *Public Health in British India: Anglo-Indian Preventive Medicine, 1859–1914*. Cambridge: Cambridge University Press, 1994.

Harvey, David. *The Condition of Postmodernity*. Oxford: Blackwell, 1989.

Hawker, Peter. *Instructions to Young Sportsmen in All That Relates to Guns and Shooting*. London: Longman, Rees, Orme, Brown, & Green, 1830.

Headley, Joel T. *The Life and Travels of General Grant*. Philadelphia: Hubbard Bros., 1879.

Hearder, H., ed. *Europe in the Nineteenth Century 1830–1880*. London: Longman, 1988.

Holland, Robert, and Diana Markides. *The British and the Hellenes: Struggles for Mastery in the Eastern Mediterranean 1850–1960*. Oxford: Oxford University Press, 2006.

Howe, Kathleen Stewart. "Mapping a Sacred Geography: Photographic Surveys by the Royal Engineers in the Holy Land, 1864–68." In *Picturing Place: Photography and the Geographical Imagination*, edited by Joan M. Schwartz and James R. Ryan, 226–42. London: I. B. Tauris, 2003.

Howell, Philip. *Geographies of Regulation: Policing Prostitution in Nineteenth-Century Britain and the Empire*. Cambridge: Cambridge University Press, 2009.

Hudson, William H. *British Birds*. London: Longmans, Green, 1895.

Hume, Edgar E. *Ornithologists of the United States Medical Army Corps*. Baltimore: Johns Hopkins University Press, 1942.

Irby, L. Howard. *British Birds: Key List*. London: R. H. Porter, 1888.

———. *The Ornithology of the Straits of Gibraltar*. London: R. H. Porter, 1875.

———. *The Ornithology of the Straits of Gibraltar*. 2nd ed. London: R. H. Porter, 1895.

Izzo, Jean-Claude, and Thierry Fabre. *La Méditerranée française*. Paris: Maisonneuve & Larose, 2000.

Jackson, Christine E., and Peter Davis. *Sir William Jardine: A Life in Natural History*. London: Leicester University Press, 2001.

Jacobs, Nancy. "Africa, Europe and the Birds between Them." In *Eco-cultural Networks and the British Empire: New Views on Environmental History*, edited by James Beattie, Edward Melillo, and Emily O'Gorman, 92–120. London: Bloomsbury, 2014.

———. *Birders of Africa: History of a Network*. New Haven, CT: Yale University Press, 2016.
James, Jude. *Lymington: An Illustrated History*. Dorset, UK: Dovecote Press, 2007.
Jardine, William. *The Naturalist's Library: Birds of Great Britain and Ireland, Incessores-rasores and Grallatores*. London: W. H. Lizars, 1839.
Jasanoff, Maya. *Edge of Empire: Conquest and Collecting in the East, 1750-1850*. London: Knopf, 2005.
Jasen, Patricia. *Wild Things: Nature, Culture, and Tourism in Ontario 1790-1914*. Toronto: University of Toronto Press, 1995.
Jones, Ted, and Anita Jones. *Fredericton and Its People*. Halifax, NS: Nimbus Publishing, 2002.
Kennedy, Dane. *The Magic Mountains: Hill Stations and the British Raj*. Berkeley: University of California Press, 1996.
Kingsley, Charles. *The Study of Natural History: A Lecture Delivered at the RA Institution, Woolwich, Oct. 3, 1871*. Woolwich, UK: Royal Artillery Institution, 1874.
Kristin, A. "Family Upupidae (Hoopoes)." In *Handbook of the Birds of the World*, vol. 6, edited by Josep del Hoyo, Andrew Elliott, and Jordi Sargatal, 396-411. Barcelona: Lynx Edicions, 2001.
Laidlaw, Zoë. *Colonial Connections 1815-45: Patronage, the Information Revolution and Colonial Government*. Manchester: Manchester University Press, 2005.
Lambert, David, and Alan Lester. *Colonial Lives across the British Empire: Imperial Careering in the Long Nineteenth Century*. Cambridge: Cambridge University Press, 2006.
Latour, Bruno. *Science in Action: How to Follow Scientists and Engineers through Society*. Cambridge, MA: Harvard University Press, 1987.
Lester, Alan. *Imperial Networks: Creating Identities in Nineteenth-Century South Africa and Britain*. London: Routledge, 2001.
———. "Introduction: New Imperial and Environmental Histories of the Indian Ocean." In *The East India Company and the Natural World*, edited by Vinita Damodaran, Anna Winterbottom, and Alan Lester, 1-15. London: Palgrave Macmillan, 2014.
Lewis, Joanna. "Southampton and the Making of an Imperial Myth: David Livingstone's Remains." In *Southampton: Gateway to the British Empire*, edited by Miles Taylor, 31-46. London: I. B. Taurus, 2007.
Livingstone, David N. *The Geographical Tradition: Episodes in the History of a Contested Enterprise*. Oxford: Blackwell, 1992.
———. *Putting Science in Its Place*. Chicago: University of Chicago Press, 2003.
———. *Science, Space, and Hermeneutics: Hettner Lecture 2001*. Heidelberg: University of Heidelberg, 2002.
Lorimer, Hayden. "Songs from Before: Shaping the Conditions for Appreciative Listening." In *Practising the Archive: Reflections on Method and Practice in Historical Geography*, edited by E. Gagen, H. Lorimer, and A. Vasudevan, 57-74. Historical Geography Research Group, Research Series, no. 40, 2007.
Lorimer, Hayden, and Katrin Lund. "Performing Facts: Finding a Way through Scotland's Mountains." In *Nature Performed: Environment, Culture and Performance*,

edited by B. Szerszynski, W. Heim, and C. Waterton, 130–44. London: Blackwells, 2003.

Lowe, Philip, Jonathan Murdoch, and Graham Cox. "A Civilised Retreat? Anti-urbanism, Rurality and the Making of an Anglo-Centric Culture." In *Managing Cities*, edited by P. Healey, S. Cameron, S. Davoudi, S. Graham, and A. Madoni-Prior, 63–83. Chichester, UK: J. Wiley, 1995.

Mackenzie, John M. *Museums and Empire: Natural History, Human Culture and Colonial Identities*. Manchester: University of Manchester Press, 2009.

Maempel, George Z. *Pioneers of Maltese Geology*. Malta: Mid-Med Bank, 1989.

Malta Garrison Library. *Second Part of the Classified Catalogue of the Malta Garrison Library, from 1 Jan 1865 to 31 Dec 1871*. Valletta, Malta: Malta Garrison Library, 1871.

Mangan, J. A., and Callum McKenzie. "The Other Side of the Coin: Victorian Masculinity, Field Sports and English Elite Education." In *Making European Masculinities: Sport, Europe, Gender*, edited by J. A. Mangan, 62–85. London: Frank Cass, 2000.

The Marquis of Ruvigny and Raineval. *Plantagenet Roll of the Blood Royal: The Isabel of Essex Volume*. Baltimore: Genealogical Publishing, 1994.

Marsden, Terry, Jonathan Murdock, Philip Lowe, Richard Munton, and Andrew Flynn. *Constructing the Countryside*. London: UCL Press, 1993.

Martin, R. Montgomery. *History of the British Possessions in the Mediterranean: Comprising Gibraltar, Malta, Gozo, and the Ionian Islands*. Vol. 7 of *The British Colonial Library*, by R. Montgomery Martin. London: Whittaker & Co., 1837.

———. *History of the Colonies of the British Empire*. London: W. H. Allen & Co., 1843.

Martins, Luciana. "Mapping Tropical Waters: British Views and Visions of Rio de Janeiro." In *Mappings*, edited by Denis Cosgrove, 148–68. London: Reaktion Books, 1999.

Mason, Adair S. *George Edwards: The Bedell and His Birds*. London: Royal College of Physicians, 1992.

Massey, Doreen. *For Space*. London: Sage, 2005.

Matless, David. *Landscape and Englishness*. London: Reaktion Books, 1998.

May, Jon, and Nigel Thrift. *TimeSpace: Geographies of Temporality*. New York: Routledge, 2001.

McCarthy, Michael. *Say Goodbye to the Cuckoo: Migratory Birds and the Impending Ecological Catastrophe*. Lanham, MD: Ivan R. Dee, 2011.

McGhie, Henry A. *Henry Dresser and Victorian Ornithology: Bird, Books and Business*. Manchester: Manchester University Press, 2017.

McGilchrist, John. *The Public Life of Queen Victoria*. New York: Cassell, Petter, & Galpin, 1868.

McGillivray, William. *A Memorial Tribute to William McGillivray*. Edinburgh: Private circulation, 1901.

McGrigor, Mary, ed. *The Scalpel and the Sword: Sir James McGrigor: The Autobiography of the Father of Army Medicine*. Dalkeith, UK: Scottish Cultural Press, 2000.

McNeill, John R. *The Mountains of the Mediterranean World*. Cambridge: Cambridge University Press, 1992.

Mearns, Barbara, and Richard Mearns. *Biographies for Birdwatchers: The Lives of Those Commemorated in Western Paleoarctic Bird Names*. London: Academic Press, 1988.

———. *The Bird Collectors*. San Diego: Academic Press, 1998.

Mitchell, Jon P. *Ambivalent Europeans: Ritual, Memory and the Public Sphere in Malta*. New York: Routledge, 2002.

Moore, D. S., A. Pandian, and J. Kosek. "The Cultural Politics of Race and Nature: Terrains of Power and Practice." In *Race, Nature, and the Politics of Difference*, edited by D. S. Moore, A. Pandian, and J. Kosek, 1–70. Durham, NC: Duke University Press, 2003.

Morris, Francis Orpen. *A History of British Birds*. London: John C. Nimmo, 1895.

Murray, John. *A Handbook for Travellers in Surrey, Hampshire, and the Isle of Wight*. 3rd ed. London: John Murray, 1876.

Myerly, Scott H. *British Military Spectacle: From the Napoleonic Wars through the Crimea*. Cambridge, MA: Harvard University Press, 1996.

Napier, Edward D. H. E. *Wild Sports in Europe, Asia and Africa*. Philadelphia: E. Ferrett, 1846.

Neilson, Mrs. Andrew. *Crimea: Its Towns, Inhabitants, and Social Customs*. London: Partridge, Oakey, & Co., 1855.

Newton, Alfred. *Manual of Zoology*. London: SPCK, 1874.

Notman, William. *Portraits of British Americans*. Montreal: William Notman, 1868.

Ó Cadhla, Stiofán. *Civilizing Ireland: Ordnance Survey 1824-1842: Ethnography, Cartography, Translation*. Dublin: Irish Academic Press, 2007.

Ogborn, Miles. *Spaces of Modernity: London's Geographies 1680-1780*. New York: Guilford Press, 1998.

Outram, Dorinda. "New Spaces in Natural History." In *Cultures of Natural History*, edited by N. Jardine, J. A. Secord, and E. C. Spary, 249–65. Cambridge: Cambridge University Press, 1996.

———. "On Being Perseus: Travel and Truth in the Enlightenment." In *Geography in the Enlightenment*, edited by C. Withers and D. Livingstone, 281–94. Chicago: University of Chicago Press, 1999.

Parry, Edward. *Memorials of Charles Parry*. London: Strahan & Co., 1870.

Pascoe, Judith. *The Hummingbird Cabinet: A Rare and Curious History of Romantic Collectors*. Ithaca, NY: Cornell University Press, 2006.

Pemble, John. *The Mediterranean Passion: Victorians and Edwardians in the South*. Oxford: Oxford University Press, 1987.

Pettitt, Clare. *Dr. Livingstone, I Presume? Missionaries, Journalists, Explorers, and Empire*. London: Profile Books, 2007.

Phillips, Richard. *Mapping Men and Empire: A Geography of Adventure*. London: Routledge, 1997.

Pick, Daniel. *Faces of Degeneration: A European Disorder, c. 1848–c. 1918*. Cambridge: Cambridge University Press, 1989.

Pindar, John. *Autobiography of a Private Soldier*. Cupar, UK: Fife News, 1877.

Playfair, Robert L. *Handbook to the Mediterranean*. London: John Murray, 1881.

Power, W. T. *Recollections of a Three Years' Residence in China; Including Peregrinations in Spain, Morocco, Egypt, India, Australia, and New Zealand*. London: Richard Bentley, 1865.

Pratt, Mary L. *Imperial Eyes: Travel Writing and Transculturation*. London: Routledge, 1992.

Raffles, Hugh. *In Amazonia: A Natural History*. Princeton, NJ: Princeton University Press, 2002.

Ramm, Agatha. "Great Britain and France in Egypt 1876–1882." In *France and Britain in Africa: Imperial Rivalry and Colonial Rule*, edited by Prosser Gifford and William Roger Louis, 73–119. New Haven, CT: Yale University Press, 1971.

Ramsay, R. G. W. *The Ornithological Works of Arthur, Ninth Marquis of Tweeddale*. London: Taylor & Francis, 1881.

Ranken, George. *Canada and the Crimea, or, Sketches of a Soldier's Life: From the Journals and Correspondence of the Late Major Ranken, R. E.* London: Longman, Green, Longman, & Roberts, 1862.

Rappaport, Helen. *Queen Victoria: A Biographical Companion*. Santa Barbara, CA: ABC-CLIO, 2003.

Rich, Adrienne. *What Is Found There: Notebooks on Poetry and Politics*. New York: Quality Paperback Book Club, 1994.

Richards, Thomas. *The Imperial Archive: Knowledge and the Fantasy of Empire*. London: Verso, 1993.

Richardson, David L. *The Anglo-Indian Passage, Homeward and Outward*. London: Madden & Malcolm, 1845.

Richardson, John, and William Swainson. *Fauna Boreali-Americana, or, the Zoology of the Northern Parts of British America*. London: John Murray, 1831.

Ritvo, Harriet. *The Animal Estate: The English and Other Creatures in the Victorian Age*. Cambridge, MA: Harvard University Press, 1987.

———. *The Platypus and the Mermaid and Other Figments of the Classifying Imagination*. Cambridge, MA: Harvard University Press, 1997.

———. "Zoological Nomenclature and the Empire of Victorian Science." In *Contexts in Victorian Science*, edited by Bernard Lightman, 334–53. Chicago: University of Chicago Press, 1997.

Robertson, William. *Journal of a Clergyman during a Visit to the Peninsula*. Edinburgh: William Blackwood, 1845.

Robinson, Ronald, and John Gallagher. *Africa and the Victorians: The Official Mind of Imperialism*. 2nd ed. London: Macmillan, 1981.

Rodger, Ella H. B. *Aberdeen Doctors: The Narrative of Medical School*. Edinburgh: W. Blackwood & Sons, 1893.

Rose, Sonya O. *Which People's War? National Identity and Citizenship in Britain 1939–1945*. Oxford: Oxford University Press, 2004.

Royal Artillery Institution. *Minutes of Proceedings of the Royal Artillery Institution*. Vol. 6. Woolwich, UK: Royal Artillery Institution, 1870.

Rupprecht, Anita. "Wonderful Adventures of Mrs. Seacole in Many Lands (1857)." In *Colonial Lives across the British Empire: Imperial Careering in the Long Nineteenth*

Century, edited by David Lambert and Alan Lester, 176–203. Cambridge: Cambridge University Press, 2006.

Ryan, James. *Picturing Empire: Photography and the Visualisation of the British Empire*. London: Reaktion Books, 1997.

Said, Edward. *Orientalism*. New York: Vintage Books, 1979.

Sala, George A. *From Waterloo to the Peninsula: Four Months' Hard Labour in Belgium, Holland, Germany, and Spain*. Vol. 2. London: Tinsley Bros., 1867.

Salvin, Osbert, and Ernst Hartert. *Catalogue of the Birds in the British Museum*. Vol. 16, *Upupæ, Trochili, and Coraciæ*, Part I. London: Trustees of the British Museum (Natural History), 1892.

Sanchez, M. G. *The Prostitutes of Serruya's Lane and Other Hidden Gibraltarian Histories*. Dewsbury, UK: Rock Scorpion Books, 2007.

Sayer, Frederick. *The History of Gibraltar and of Its Political Relation to Events in Europe*. London: Chapman & Hall, 1865.

Schieffelin, Edward L. "Problematizing Performance." In *Ritual, Performance, Media*, edited by Felicia Hughes-Freeland, 195–209. London: Routledge, 1998.

Schwartz, Joan M. "Photographs from the Edge of Empire." In *Cultural Geography in Practice*, edited by A. Blunt, P. Gruffudd, J. May, M. Ogborn, and D. Pinder, 154–71. New York: Oxford University Press, 2003.

Schwartz, Joan M., and James R. Ryan, eds. *Picturing Place: Photography and the Geographical Imagination*. London: I. B. Taurus, 2003.

Sclater, Philip L. *List of the Vertebrated Animals Now or Lately Living in the Gardens of the Zoological Society of London*. 5th ed. London: Longmans, Green, Reader, & Dyer, 1877.

Seacole, Mary. *Wonderful Adventures of Mrs. Seacole in Many Lands*. London: James Blackwood Paternoster Row, 1857.

Shapin, Steven. *A Social History of Truth: Civility and Science in Seventeenth-Century England*. Chicago: University of Chicago Press, 1994.

Sharp, Joanne P. "Gendering Nationhood: A Feminist Engagement with National Identity." In *BodyScape: Destabilising Geographies of Gender and Sexuality*, edited by Nancy Duncan, 97–108. London: Routledge, 1996.

Sharpe, Richard B. *Hand-Book to the Birds of Great Britain*. Vol. 4. London: Wyman & Sons, 1897.

Shelley, George E. *A Handbook to the Birds of Egypt*. London: John Van Voorst, 1873.

Smyth, John. *Sandhurst: The History of the Royal Military Academy*. London: Weidenfeld & Nicolson, 1961.

Snow, David W., and Christopher W. Perrins. *The Birds of the Western Palearctic*. Vol. 1, *Non-passerines*. Oxford: Oxford University Press, 1998.

Srhir, Khalid Ben, Malcolm Williams, and Gavin Waterson. *Britain and Morocco during the Embassy of John Drummond-Hay, 1845–1886*. New York: Routledge, 2005.

Stagl, Justin. *A History of Curiosity: The Theory of Travel 1550–1800*. London: Harwood Academic Publishers, 1995.

Stanley, Peter. *White Mutiny: British Military Culture in India*. New York: New York University Press, 1998.

Stevenson, Henry. *The Birds of Norfolk, with Remarks on Their Habits, Migration, and Local Distribution*. London: John Van Voorst, 1866.
Streseman, E. *Ornithology from Aristotle to the Present*. Cambridge, MA: Harvard University Press, 1975.
St. Thomas and All Saints. *St. Thomas and All Saints: Lymington*. Lymington, UK: St. Thomas and All Saints, n.d.
Tasker, Meg. *"Struggle and Storm": The Life and Death of Francis Adams*. Melbourne: Melbourne University Press, 2001.
Taylor, Miles. *Southampton: Gateway to the British Empire*. London: I. B. Taurus, 2007.
Thackeray, William M. *Rock of Empire: Literary Visions of Gibraltar, 1700-1900*. Edited by M. G. Sanchez. Gibraltar: Gibraltar Chronicle, 2001.
Thomas, Ronald R. "The Home of Time: The Prime Meridian, the Dome of the Millennium, and Postnational Space." In *Nineteenth-Century Geographies: The Transformation of Space from the Victorian Age to the American Century*, edited by Helena Michie and Ronald R. Thomas, 23-39. New Brunswick, NJ: Rutgers University Press, 2003.
Timbers, Ken, ed. *The Royal Artillery, Woolwich: A Celebration*. London: Royal Artillery and Third Millennium Publishing, 2008.
Tippett, Brian. "W. H. Hudson in Hampshire." In *Hampshire Papers* 27. Winchester, UK: Hampshire County Council, 2004.
Verner, Willoughby. *My Life among the Wild Birds of Spain*. London: Whiterby, 1909.
Vickers, P. H. *"A Gift So Graciously Bestowed": The History of the Prince Consort's Library*. Winchester, UK: Hampshire County Council, 1992.
Wallace, Alfred R. *The Geographical Distribution of Animals*. London: MacMillan, 1876.
Wallerstein, Immanuel. *Unthinking Social Science: The Limits of Nineteenth-Century Paradigms*. 2nd ed. Philadelphia: Temple University Press, 1991.
Walters, Michael. *A Concise History of Ornithology*. New Haven, CT: Yale University Press, 2003.
Warr, Frances E. *Manuscripts and Drawings in the Ornithology and Rothschild Libraries of the Natural History Museum at Tring*. Tring, UK: British Ornithologists' Club in association with the Natural History Museum, London, 1996.
Whatmore, Sarah. *Hybrid Geographies: Nature, Cultures, Spaces*. London: Sage, 2002.
White, Gilbert. *The Natural History of Selborne*. New York: Penguin Books, 1977.
Williams, Brian G. *The Crimean Tatars: The Diaspora Experience and the Forging of a Nation*. Leiden: Brill, 2001.
Wilson, Robert M. *Seeking Refuge: Birds and Landscapes of the Pacific Flyway*. Seattle: University of Washington Press, 2010.
Withers, Charles W. J. *Geography, Science and National Identity: Scotland since 1520*. Cambridge: Cambridge University Press, 2001.
Wollaston, Alexander F. R. *Life of Alfred Newton: Professor of Comparative Anatomy, Cambridge University, 1866-1907*. London: John Murray, 1921.
Wolseley, Garnet. *Soldier's Pocket-Book for Field Service*. London: MacMillan & Co., 1871.
Woodward, Rachel. *Military Geographies*. Oxford: Blackwell, 2004.

Woollacott, Angela. *Gender and Empire*. Basingstoke, UK: Palgrave Macmillan, 2006.
Wynn, Graeme. "Foreword." In *Hunting for Empire: Narratives of Sport in Rupert's Land, 1840-70*, edited by Greg Gillespie, xi-xx. Vancouver: University of British Columbia Press, 2008.
Yarrell, William. *A History of British Birds*. London: John Van Voorst, 1843.
Yarrell, William, Alfred Newton, and Howard Saunders. *A History of British Birds*. 4th ed. London: John Van Voorst, 1874.

Periodicals

Adams, Andrew L. "The Birds of Cashmere and Ladakh." *Proceedings of the Zoological Society of London* 26 (1858): 169-90.
———. "Migrations of European Birds." *Popular Science Review* 4 (1865): 324-34.
———. "Notes on Certain Meteorological Phenomena in Connexion with Cholera and Other Diseases." *Medical Times and Gazette* 1 (1867): 306.
———. "Notes on the Habits, Haunts, etc. of Some of the Birds of India." *Proceedings of the Zoological Society of London* 26 (1858): 466-512.
Alberti, Samuel J. M. M. "Constructing Nature behind Glass." *Museum and Society* 6, no. 2 (July 2008): 73-97.
———. "Objects and the Museum." *Isis* 96 (2005): 559-71.
Alerstam, Thomas, M. Hake, and Nils Kjellén. "Temporal and Spatial Patterns of Repeated Migratory Journeys by Ospreys." *Animal Behaviour* 71, no. 3 (2006): 555-66.
Alonso, J. C. "The Great Bustard: Past, Present and Future of a Globally Threatened Species." *Ornis Hungarica* 22 (2014): 1-13.
Andryushchenko, Y. A., and V. M. Popenko. "Birds and Power Lines in Steppe Crimea: Positive and Negative Impacts." Пернатые хищники и их охрана 24 (2012): 34-41.
Anon. "Additions to the Zoological Society's Gardens." *Nature* 7 (1873): 371.
———. "Crimean Heroes and Trophies." *Illustrated London News*, 12 April 1856, 369.
———. "Crimean Snowdrop." *Illustrated London News*, 5 April 1856, 359.
———. "Gibraltar and Neighbourhood." *Scientific American Supplement* 711 (17 August 1889): 11352.
———. "How I Spent My Five Weeks' Leave." *Freemasons' Magazine and Masonic Mirror*, 27 November 1869, 427.
———. "Irby's Birds of Gibraltar."*Nature* 12 (December 1875): 364-65.
———. "Museum and Library." *Journal of the Royal Artillery* 1 (1858): 367, 441.
———. "Notes on Gibraltar." *Hogg's Weekly Instructor* 108 (March 1847): 51-54.
———. "Notices of Serials." *Natural History Review* 4 (1857): 57.
———. "Obituary: Andrew Leith Adams." *British Medical Journal* 2 (19 August 1882): 338.
———. "Obituary: Captain George Ernest Shelley, FZS." *Journal of the South African Ornithologists' Union* 7 (1911): 93-94. Reprinted from *Ibis*, April 1911.
———. "Obituary: Philip Savile Grey Reid." *Ibis* 3 (1915): 365-67.
———. "Review of Mr. Bree's 'Birds of Europe Not Observed in the British Isles.'" *Ibis* 1 (1859): 88.

———. "Russian Guns and Bells from Sebastopol, Just Received at Woolwich Arsenal." *Illustrated London News*, 23 February 1856, 209.
———. "Taxidermy." *Journal of the Royal Artillery* 2 (1861): 424.
———. "To the Editor of the United Service Journal." *United Service Magazine* (1830): 367–68.
Arnold, Sergeant. "Report on the Climate of Aldershot Camp." *Meeting of the British Association for the Advancement of Science* 36 (1867): 15.
Bailey, Harry P. "Toward a Unified Concept of the Temperate Climate." *Geographical Review* 54, no. 4 (October 1964): 516–45.
Baldacchino, G. "A Nationless State? Malta, National Identity and the EU." *West European Politics* 25, no. 4 (October 2002): 191–206.
Barnes, Trevor J. "Lives Lived and Lives Told: Biographies of Geography's Quantitative Revolution." *Environment and Planning D: Society and Space* 19, no. 4 (2001): 409–29.
Becher, E. F. "Zoological Notes from Gibraltar." *Zoologist* 3 (1883): 100.
Ben-Artzi, Yossi. "The Idea of a Mediterranean Region in Nineteenth- to Mid-Twentieth-Century German Geography." *Mediterranean Historical Review* 19, no. 2 (2004): 2–15.
Bhabha, Homi K. "The Other Question . . . Homi K. Bhabha Reconsiders the Stereotype and Colonial Discourse." *Screen* 24, no. 6 (1983): 18–36.
Bird, James. "The Acclimation of European Troops for Service in India, and the Organization There Most Suited to Secure Efficiency in the Field." *Journal of the Royal United Service Institution* 3 (1860): 324–36.
Blakiston, Thomas W. "Birds of the Crimea." *Zoologist* 15 (1857): 5348–680.
———. "Interior of British North America." *Ibis* 5 (1863): 141.
Blouet, Olwyn M. "Sir William Reid, F.R.S., 1791–1858: Governor of Bermuda, Barbados and Malta." *Notes and Records of the Royal Society* 40 (1986): 169–91.
Blunt, Alison. "Imperial Geographies of Home: British Domesticity in India, 1886–1925." *Transactions of the Institute of British Geographers* 24 (1999): 421–40.
Bonhomme, Brian. "Nested Interests: Assessing Britain's Wild-Bird Protection Laws, 1869–1880." *Nineteenth-Century Studies* 19 (2005): 47–68.
Boucher, Nancy B., and Ken Cruikshank. "Sportsmen and Pothunters: Environment, Conservation, and Class in the Fishery of Hamilton Harbour, 1858–1914." *Sport History Review* 28, no. 1 (May 1997): 1–18.
Braun, Bruce. "Producing Vertical Territory: Geology and Governmentality in Late Victorian Canada." *Ecumene* 7, no. 1 (2000): 7–46.
Browne, Janet. "A Science of Empire: British Biogeography before Darwin." *Revue d'Histoire des Sciences* 45, no. 4 (1992): 453–75.
Burant, Jim. "The Military Artist and the Documentary Art Record." *Archivaria* 26 (1988): 33–51.
———. "Record of Empire, 1835–1896: The John A. Vesey Kirkland Album." *Archivaria* 22 (1988): 120–28.
Burke, S. D. A., and Larry Sawchuk. "Alien Encounters: The Jus Soli and Reproductive Politics in the 19th-Century Fortress and Colony of Gibraltar." *History of the Family* 4, no. 6 (2001): 531–61.

Burnside, Robert J., Zsolt Végvári, Richard James, and Sandor Konyhás. "Human Disturbance and Conspecifics Influence Display Site Selection by Great Bustards *Otis tarda*." *Bird Conservation International* 24, no. 1 (2014): 32-44.

Butzer, Karl. "Environmental History in the Mediterranean World: Cross-Disciplinary Investigation of Cause-and-Effect for Degradation and Soil Erosion." *Journal of Archaeological Science* 32 (2005): 1773-800.

Camerini, Jane R. "Evolution, Biogeography, and Maps: An Early History of Wallace's Line." *Isis* 84 (1993): 700-727.

Cameron, Laura. "Distinguished Historical Geography Lecture 2010: Digging in the Dirt: Unnatural Histories and the 'Art of Not Dividing.'" *Historical Geography* 38 (2010): 5-22.

———. "Oral History in the Freud Archives: Incidents, Ethics and Relations." *Historical Geography* 29 (2001): 38-44.

Cameron, Laura, and David Matless. "Benign Ecology: Marietta Pallis and the Floating Fen of the Delta of the Danube, 1912-1916." *Cultural Geographies* 10 (2003): 253-77.

———. "Translocal Ecologies: The Norfolk Broads, the 'Natural,' and the International Phytogeographical Excursion, 1911." *Journal of the History of Biology* (published online July 2010): 1-27.

Carte, Alexander. "Donations to the Royal Dublin Society." *Journal of the Royal Dublin Society* 1 (1856): 286.

———. "Report on the State and Progress of the Museum of Natural History of the Royal Dublin Society for the Year 1855." *Journal of the Royal Dublin Society* 1 (1858): 40.

Chadha, Ashish. "Ambivalent Heritage: Between Affect and Ideology in a Colonial Cemetery." *Journal of Material Culture* 11 (2006): 339-63.

Christopher, A. J. "The Quest for a Census of the British Empire c. 1840-1940." *Journal of Historical Geography* 34 (2008): 268-85.

Clarke, Hyde. "The Military Advantages of a Daily Mail Route to India through Turkey and the Persian Gulf." *Journal of the Royal United Service Institution* 12 (1869): 181-91.

Claval, Paul. "About Rural Landscapes: The Invention of the Mediterranean and the French School of Geography." *Erde* 138, no. 1 (2007): 7-24.

Constantine, Stephen. "Monarchy and Constructing Identity in 'British' Gibraltar, c. 1800 to the Present." *Journal of Imperial and Commonwealth History* 1, no. 34 (2006): 23-44.

Dalton, Canon. "The Colonial Conference 1887." *Proceedings of the Royal Colonial Institute* 19 (1888): 32-33.

Darwin, John. "Imperialism and the Victorians: The Dynamics of Territorial Expansion." *English Historical Review* 112, no. 447 (1997): 614-42.

Datson, Lorraine. "Type Specimens and Scientific Memory." *Critical Inquiry* 31, no. 1 (2004): 153-82.

Davies, Thomas. "On a Method of Preparing Birds for Preservation." *Philosophical Transactions of the Royal Society from 1770 to 1776* 14 (1809): 34-35.

Denny, N. D. "British Temperance Reformers and the Island of Malta 1815-1914." *Melita Historica* 9 (1987): 329-45.

DeSilvey, Caitlin. "Observed Decay: Telling Stories with Mutable Things." *Journal of Material Culture* 11 (2006): 318–38.
Dodds, Klaus, David Lambert, and Bridget Robison. "Loyalty and Royalty: Gibraltar, the 1953–54 Royal Tour and the Geopolitics of the Iberian Peninsula." *Twentieth Century British History* 18, no. 3 (2007): 365–90.
Driver, Felix. "Sub-merged Identities: Familiar and Unfamiliar Histories." *Transactions of the Institute of British Geographers* 20 (1995): 410–13.
Drummond Hay, Maurice. "Occurrence of the Hoopoe in the Tay District." *Transactions and Proceedings of the Perthshire Society of Natural Science* 1–2 (1893): cvlvii–cvlviii.
Duncan, James S. "The Struggle to Be Temperate: Climate and 'Moral Masculinity' in Mid-Nineteenth Century Ceylon." *Singapore Journal of Tropical Geography* 21, no. 1 (2000): 34–47.
Edney, Matthew H. "British Military Education, Mapmaking and Military Mindedness in the Late Enlightenment." *Cartographic Journal* 31, no. 1 (1994): 14–20.
Erzini, Nadia. "Hal Yaslah Li-Taqansut (Is He Suitable for Consulship?): The Moroccan Consuls in Gibraltar during the Nineteenth Century." *Journal of North African Studies* 12, no. 4 (2007): 517–29.
Farber, Paul L. "The Development of Taxidermy and the History of Ornithology." *Isis* 68 (1977): 350–566.
Febvre, Lucien. "The Mediterranean Is the Sum of Its Routes." *Annales d'Histoire Sociale* (11 January 1940): 70.
Feilden, H. W. "Ospreys in Hampshire." *Zoologist* 7 (1872): 2991–92.
Finlayson, Clive. "William Willoughby Cole Verner." *Gibraltar Heritage Journal* 3 (1996): 91–99.
Finnegan, Diarmid A. "'An Aid to Mental Health': Natural History, Alienists and Therapeutics in Victorian Scotland." *Studies in History and Philosophy of Science Part C: Studies in History and Philosophy of Biological and Biomedical Sciences* 39, no. 3 (2008): 326–37.
Fletcher, Alison A. "'Mother Seacole': Victorian Domesticity on the Battlefields of the Crimean War." *Minerva Journal of Women and War* 1, no. 2 (2007): 7–21.
Forbes, Thomas R. "William Yarrell, British Naturalist." *Proceedings of the American Philosophical Society* 106, no. 6 (1962): 505–15.
Ford, John M. T. "Francis Adams (1796–1861): Aberdeen, UK." *Journal of Medical Biography* 16, no. 1 (2008): 56.
Forsyth, Isla. "On the Edges of Military Mobilities." *Political Geography* 56 (2017): 48–50.
———. "The Practice and Poetics of Fieldwork: Hugh Cott and the Study of Camouflage." *Journal of Historical Geography* 43 (2014): 128–37.
Foster, Paul. "The Gibraltar Collections: Gilbert White (1720–1793) and John White (1727–1780), and the Naturalist and Author Giovanni Antonio Scopoli (1723–1788)." *Archives of Natural History* 34, no. 1 (2007): 30–46.
Frome, Major General. "Moncrieff's System of Artillery." *Papers on Subjects Connected with the Duties of the Corps of the Royal Engineers* 18 (1870): 33.

Greer, Kirsten. "Geopolitics and the Avian Imperial Archive: The Zoogeography of Region-Making in the Late 19th-Century British Mediterranean." *Annals of the Association of American Geographers* 103 (2013): 1317–31.

———. "Placing Colonial Ornithology: Imperial Ambiguities in Upper Canada, 1791–1841." *Scientia Canadensis* 31, nos. 1–2 (2008): 85–112.

———. "Untangling the Avian Imperial Archive." *Antennae: The Journal of Nature in Visual Culture*. Special issue on Alternative Ornithologies 20 (2012): 59–71.

———. "Zoogeography and Imperial Defence: Tracing the Contours of the Nearctic Region in the Temperate North Atlantic, 1838–1880s." *Geoforum* 65 (2015): 454–64.

Greer, Kirsten, and Sonje Bols. "'She of the Loghouse Nest': Gendering Historical Ecological Reconstructions in Northern Ontario." *Historical Geography* 44 (2016): 45–67.

Greer, Kirsten, and Laura Cameron. "The Use and Abuse of Ecological Constructs." *Geoforum* 65 (2015): 451–53.

Gregory, Derek. "Between the Book and the Lamp: Imaginative Geographies of Egypt, 1849–50." *Transactions of the Institute of British Geographers* 20 (1995): 29–57.

Griesemer, James R. "Modeling in the Museum: On the Role of Remnant Models in the Work of Joseph Grinnell." *Biology and Philosophy* 5 (1990): 3–36.

Haraway, Donna. "Teddy Bear Patriarchy: Taxidermy in the Garden of Eden, New York City, 1908–36." *Social Text* 11 (Winter 1984/1985): 19–64.

Harley, J. B. "Historical Geography and the Cartographic Illusion." *Journal of Historical Geography* 15 (1989): 80–91.

Hawker, William H. "Zoology from the Seat of War." *Zoologist* 14 (1856): 5203.

Hevly, Bruce. "Heroic Science of Glacier Motion." *Osiris*, 2nd ser., Science in the Field 11 (1996): 66–86.

Holling, M., and the Rare Breeding Birds Panel. "Rare Breeding Birds in the United Kingdom in 2010." *British Birds* 105 (2012): 352–416.

Homans, Margaret. "'To the Queen's Private Apartments': Royal Family Portraiture and the Construction of Victoria's Sovereign Obedience." *Victorian Studies* 37, no. 1 (1993): 1–41.

Howell, Philip. "Sexuality, Sovereignty and Space: Law, Government and the Geography of Prostitution in Colonial Gibraltar." *Social History* 29 (2004): 444–64.

Hunt, James. "On Ethno-climatology; or, the Acclimatization of Man." *Report of the Annual Meeting* 31 (1862): 136.

Ingold, Tim. "Rethinking the Animate, Re-animating Thought." *Ethnos* 71 (2006): 9–20.

Irby, L. H. "Lists of Birds Observed in the Crimea." *Zoologist* 15 (1857): 5360.

———. "Notes of Birds Observed in Oudh and Kumaon." *Ibis* 3 (1861): 217–51.

———. "Notes on the Birds of the Straits of Gibraltar." *Ibis* 21 (1879): 342–46.

Jacobs, Nancy. "The Intimate Politics of Ornithology in Colonial Africa." *Comparative Studies in Society and History* 48 (2006): 564–603.

Jankovic, Vladimir. "The Last Resort: A British Perspective on the Medical South, 1815–1870." *Journal of Intercultural Studies* 27, no. 3 (2006): 271–98.

Jardine, William. "Hints for Preparing and Transmitting Ornithological Specimens from Foreign Countries." *Contributions to Ornithology, 1848* (1849): 3–12.

Johnson, Kristin. "Type-Specimens of Birds as Sources for the History of Ornithology." *Journal of the History of Collections* 17, no. 2 (2005): 173–88.
Johnson, Nuala. "Cast in Stone: Monuments, Geography, and Nationalism." *Environment and Planning D: Society and Space* 13, no. 1 (1995): 51–65.
Kelham, Henry R. "Ornithological Notes Made in the Straits Settlements and in the Western States of the Malay Peninsula." *Ibis* 23 (1881): 362–95.
Kelsall, J. E. "A Briefly Annotated List of the Birds of Hampshire and the Isle of Wight." *Papers and Proceedings of the Hampshire Field Club* 1 (1887): 90–122.
Kessler, A. E., N. Batbayar, T. Natsagdorj, D. Batsuur, and A. T. Smith. "Satellite Telemetry Reveals Long-Distance Migration in the Asian Great Bustard Otis tarda dybowskii." *Journal of Avian Biology* 44 (2013): 311–20.
Kohout, Amy. "More Than Birds: Loss and Reconnection at the National Museum of Natural History." *Museum History Journal* 10 (2017): 83–96.
Kupperman, Karen O. "Fear of Hot Climates in the Anglo-American Colonial Experience." *William and Mary Quarterly* 41 (April 1984): 213–40.
Lambert, David. "'As Solid as the Rock?' Place, Belonging and the Local Appropriation of Imperial Discourse in Gibraltar." *Transactions of the Institute of British Geographers* 30 (2005): 206–20.
Lathbury, G. "A Review of the Birds of Gibraltar and Its Surrounding Waters." *Ibis* 112 (1970): 25.
Lester, Alan. "Commentary: New Directions for Historical Geographies of Colonialism." *New Zealand Geographer* 71, no. 3 (2015): 120–23.
———. "Imperial Circuits and Networks: Geographies of the British Empire." *History Compass* 4, no. 1 (2005): 124–41.
Lipscomb, Susan B. "Introducing Gilbert White: An Exemplary Natural Historian and His Editors." *Victorian Literature and Culture* 35 (2007): 551–67.
Livingstone, David N. "Tropical Climate and Moral Hygiene: The Anatomy of a Victorian Debate." *British Journal for the History of Science* 32, no. 1 (March 1999): 93–110.
Lorimer, Hayden. "Herding Memories of Humans and Animals." *Environment and Planning D: Society and Space* 24 (2006): 497–518.
MacDonald, Helen. "What Makes You a Scientist Is the Way You Look at Things: Ornithology and the Observer 1930–1955." *Studies in History and Philosophy of Science Part C* 33 (2002): 53–77.
MacLeod, Roy. "'Strictly for the Birds': Science, the Military and the Smithsonian's Pacific Ocean Biological Survey Program, 1963–1970." *Journal of the History of Biology* 34, no. 2 (June 2001): 315–52.
Maempel, George Z. "T. A. B. Spratt (1811–88) and His Contribution to Maltese Geology." *Melita Historica* 9 (1986): 271–308.
Manai, Adel. "Anglo-Tunisian Commercial Relations in the Nineteenth Century: A Travel Note." *Journal of North African Studies* 11, no. 4 (2006): 365–72.
Matless, David. "Moral Geographies of English Landscape." *Landscape Research* 22, no. 2 (July 1997): 141–55.
Maxwell, Herbert. "Birds." *Nineteenth Century* 28 (1890): 914–26.

McGhie, Henry A. "Contextual Research and the Postcolonial Museum: The Example of Henry Dresser." In *Annalen des Naturhistorischen Museums Wien* (Proceedings of the Fifth International Conference of European Bird Curators), 49–65. Vienna: Natural History Museum, 2010.

Mills, Sara. "Cultural-Historical Geographies of the Archive: Fragments, Objects and Ghosts." *Geography Compass* 7, no. 10 (October 2013): 701–13.

Moore-Colyer, R. J. "Feathered Women and Persecuted Birds: The Struggle against the Plumage Trade, c. 1860–1922." *Rural History* 11 (2000): 57–73.

More, A. G. "On the Distribution of Birds of Great Britain during the Nesting-Season." *Ibis* 1 (1865): 1–27.

Moreau, R. E. "Centenarian *Ibis*." *Ibis* 101 (1959): 19–38.

Morin, Karen. "Charles P. Daly's Gendered Geography, 1860–1890." *Annals of the Association of American Geographers* 98, no. 4 (December 2008): 897–919.

———. "Embodying Tropicalities: Commentary on Felix Driver's 'Imagining the Tropics: Views & Visions of the Tropical World.'" *Singapore Journal of Tropical Geography* 25, no. 1 (2004): 23–25.

Morin, Karen M. "Surveying Britain's Informal Empire: Rose Kingsley's 1872 Reconnaissance for the Mexican National Railway." *Journal of Historical Geography* 89, no. 3 (1999): 489–514.

Morrison, James H. "Soldiers, Storms and Seasons: Weather Watching in Nineteenth Century Halifax." *Nova Scotia Historical Quarterly Society* 10, nos. 3–4 (September–December 1980): 224–25.

Nash, Catherine. "Performativity in Practice: Some Recent Work in Cultural Geography." *Progress in Human Geography* 24 (2000): 653–64.

Neumann, Roderick P. "Dukes, Earls, and Ersatz Edens: Aristocratic Nature Preservationists in Colonial Africa." *Environment and Planning D: Society and Space* 14, no. 1 (1996): 79–98.

———. "Moral and Discursive Geographies in the War for Biodiversity in Africa." *Political Geography* 23, no. 7 (2004): 813–37.

Newton, Alfred. "On the Possibility of Taking an Ornithological Census." *Ibis* 3 (1861): 190–96.

Ogborn, Miles, and Chris Philo. "Soldiers, Sailors and Moral Locations in Nineteenth-Century Portsmouth." *Area* 26, no. 3 (1994): 221–31.

Okihiro, Gary Y. "Unsettling the Imperial Sciences." *Environment and Planning D: Society and Space* 28, no. 5 (2010): 745–58.

Ophir, Adi, and Stephen Shapin. "The Place of Knowledge: A Methodological Survey." *Science in Context* 4 (1991): 3–22.

Paasi, Anssi. "Territorial Identities as Social Constructs." *Hagar* 1 (2000): 91–113.

Parenteau, Bill. "Angling, Hunting and the Development of Tourism in Late Nineteenth Century Canada: A Glimpse at the Documentary Record." *Archivist* 177 (1998): 10–19.

Patchett, Merle. "Taxidermy Workshops: Differently Figuring the Working of Bodies and Bodies at Work in the Past." *Transactions of the Institute of British Geographers* 42, no. 3 (September 2017): 390–404.

Patchett, Merle, and Kate Foster. "Repair Work: Surfacing the Geographies of Dead Animals." *Museum and Society* 6, no. 2 (2008): 98–122.

Peers, Douglas M. "Colonial Knowledge and the Military in India, 1780–1860." *Journal of Imperial and Commonwealth History* 33 (2005): 157–80.

———. "Privates off Parade: Regimenting Sexuality in the Nineteenth-Century Indian Empire." *International History Review* 20, no. 4 (1998): 823–54.

———. "Soldiers, Surgeons and the Campaigns to Combat Sexually Transmitted Diseases in Colonial India, 1805–1860." *Medical History* 42 (1998): 137–60.

Perera, J. B. "The Language of Exclusion in F. Solly Flood's 'History of the Permit System in Gibraltar.'" *Journal of Historical Sociology* 3, no. 20 (2007): 209–34.

Pieters, F. J. M. "Notes on the Menagerie and Zoological Cabinet of Stafholder William V of Holland, Directed by Aernout Vosmaer." *Journal for the Bibliography of Natural History* 9 (April 1980): 451–542.

Poliquin, Rachel. "The Beastly Art of Taxidermy." *TREK*, Spring/Summer 2011, 11–16.

Portlock, Joseph E. "On the Advantage of Cultivating the Natural and Experimental Sciences, as Promoting Social Comfort and Practical Utility of Military Men." *Royal United Service Institution* 3 (1860): 290–306.

Quintero, Camilo. "Trading in Birds: Imperial Power, National Pride, and the Place of Nature in U.S.-Colombia Relations." *Isis* 102, no. 3 (2011): 421–45.

Raffles, Hugh. "Local Theory: Nature and the Making of an Amazonian Place." *Cultural Anthropology* 14, no. 3 (1999): 323–60.

———. "The Uses of Butterflies." *American Ethnologist* 28, no. 3 (August 2001): 513–48.

Rawlinson, H. C. "The Military Advantages of a Daily Mail-Route to India through Turkey and the Persian Gulf." *Royal United Service Institution* 12 (1869): 181–91.

Reichlin, Thomas, Michael Schaub, Myles H. M. Menz, Muriele Mermod, Patricia Portner, Raphaël Arlettaz, and Lukas Jenni. "Migration Patterns of *Hoopoe Upupa epops* and *Wryneck Jynx torquilla*: An Analysis of European Ring Recoveries." *Journal of Ornithology* 150 (2009): 393–400.

Reid, Philip S. G. "The Birds of the Bermudas." *Royal Gazette* (1883): 1–43.

———. "Winter Notes from Morocco." *Ibis* 3 (1885): 241–55.

Rome, Adam. "Nature Wars, Culture Wars: Immigration and Environmental Reform in the Progressive Era." *Environmental History* 13, no. 3 (2008): 432–53.

Ross, W. A. "The Cultivation of Scientific Knowledge by Regimental Officers of the British Army." *Royal United Service Institution* 16 (1873): 774–81.

Russell, Douglas G. D., Mike Hansell, and Maggie Reilly. "Constructive Behaviour: Developing a UK Nest Collection Resource." Paper presented at "Collections in Context," 5th International Meeting of European Bird Curators, Natural History Museum, Vienna, 29 August 2007.

Sanchez, M. G., ed. *Rock of Empire: Literary Visions of Gibraltar, 1700–1900*. Gibraltar: Gibraltar Chronicle, 2001.

Schwartz, Joan M. "Reading Robin Kelsey's *Archive Style* across the Archival Divide." *Journal of Archival Organization* 6, no. 3 (2008): 201–10.

———. "William Notman's Hunting Photographs, 1866." *Archivist* 117 (1998): 20–29.
Scicluna, Sandra, and Paul Knepper. "Prisoners of the Sun: The British Empire and Imprisonment in Malta in the Early Nineteenth Century." *British Journal of Criminology* 48, no. 4 (2008): 502–21.
Sclater, Philip L. "On the General Geographical Distribution of the Members of the Class Aves." *Journal of the Proceedings of the Linnean Society: Zoology* 2 (1858): 130–45.
S. D. W. "ART. II. *On the Habits, Haunts, and Nidification of the Robin Redbreast (Rubecu/afamiliaris* Blyth)." *Magazine of Natural History* 9 (1836): 6.
Sharpe, Richard B. "Ornithology at South Kensington." *English Illustrated Magazine* 5 (1888): 166–75.
Skinner, J. "British Constructions with Constitutions: The Formal and Informal Nature of 'Island' Relations on Montserrat and Gibraltar." *Social Identities* 8, no. 2 (June 2002): 301–20.
Smurthwaite, Henry. "Notes on the Great Bustard." *Zoologist* 15 (1857): 5517–18.
Smyth, W. H. "To the Editor of the Journal." *United Service Magazine* (1829): 625.
Star, Susan L., and James R. Griesemer. "Institutional Ecology, 'Translations' and Boundary Objects: Amateurs and Professionals in Berkeley's Museum of Vertebrate Zoology." *Social Studies of Science* 19 (1987): 387–420.
Strickland, Hugh E. "On the Occurrence of *Charadrius Virginiacus* (Borkh) at Malta." *Annals and Magazine of Natural History* 5 (1850): 40.
Summers, Anne. "Pride and Prejudice: Ladies and Nurses in the Crimean War." *History Workshop Journal* 16, no. 1 (1983): 33–56.
Taylor, George C. "Ornithological Observations in the Crimea, Turkey, Sea of Azov, and Crete, during the Years 1854–55." *Ibis* 14 (1872): 224–37.
Tebbutt, Melanie. "Rambling and Manly Identity in Derbyshire's Dark Peak, 1880s–1920s." *Historical Journal* 49, no. 4 (2006): 1125–53.
Thomson, William. "Notice of Migratory Birds Which Alighted on, or Were Seen from, *HMS Beacon*, Capt. Graves, on the Passage from Malta to the Morea at the End of April 1841." *Annals and Magazine of Natural History* 8 (1842): 125.
Tivers, J. "'The Home of the British Army': The Iconic Construction of Military Defence Landscapes." *Landscape Research* 24, no. 3 (November 1999): 303–19.
Verner, Willoughby. "Obituary." *Ibis* 47 (July 1905): 501–5.
Wallace, Alfred R. "Letter from Mr. Wallace Concerning the Geographical Distribution of Birds." *Ibis* 1 (1859): 449.
Watkins, Charles W. "The Ornithology of Andalusia." *Zoologist* 14 (1856): 5315.
Webster, Michael S., Peter Marra, Susan Haig, and Staffan Bensch. "Links between Worlds: Unravelling Migratory Connectivity." *TRENDS in Ecology and Evolution* 17, no. 2 (February 2002): 76–83.
Wilford, E. N. "Proceedings of a General Meeting, Held on Friday 20 June 1856." *Journal of the Royal Artillery* 1 (1858): 363–78.
Willis, J. R. "List of Birds of Nova Scotia. Compiled from Notes by Lieutenant Blakiston, R.A., and Lieutenant Bland, R.E., Made in 1852–1855, by Professor J. R. Willis, of Halifax." In *Smithsonian Institution Annual Report for 1858*, 280–86. Washington, DC: James B. Steedman, 1859.

Wilson, Robert M. "Directing the Flow: Migratory Waterfowl, Scale and Mobility in Western North America." *Environmental History* 7, no. 2 (April 2002): 247–66.
Wright, Charles A. "List of the Birds Observed in the Islands of Malta and Gozo." *Ibis* 3 (1864): 42–73.
Young, Charles M. "Aristotle on Temperance." *Philosophical Review* 97, no. 4 (October 1988): 521–42.
Zeller, Suzanne. "Classical Codes: Biogeographical Assessments of Environment in Victorian Canada." *Journal of Historical Geography* 24 (2008): 20–35.
———. "Humboldt and the Habitability of Canada's Great Northwest." *Geographical Review* 96, no. 3 (July 2006): 382–98.

Electronic Sources

Anon. "New Protection for Migratory Birds and Their 'Flyways.'" *Ecologist*, 14 November 2014. http://www.theecologist.org/News/news_round_up/2630666/new_protection_for_migratory_birds_and_their_flyways.html.
BBC. "Ospreys Return to Northumberland from Africa." *BBC News*, 18 April 2011. http://www.bbc.com/news/uk-england-tyne-13117691.
BirdLife International. "BirdLife Malta Raptor Camp — September 2010 Video." 17 September 2010. Video, 2:17. https://www.birdlife.org/europe-and-central-asia/news/birdlife-malta-raptor-camp-september-2010-video.
———. "*Oriolus oriolus*." *BirdLife International IUCN Red List of Threatened Species* (2016). http://datazone.birdlife.org/species/factsheet/eurasian-golden-oriole-oriolus-oriolus. Accessed 22 August 2017.
———. "*Pandion haliaetus*." *BirdLife International IUCN Red List of Threatened Species* (2015). http://datazone.birdlife.org/species/factsheet/osprey-pandion-haliaetus. Accessed 23 August 2017.
———. "*Upupa epops*." *BirdLife International IUCN Red List of Threatened Species* (2016). http://datazone.birdlife.org/species/factsheet/common-hoopoe-upupa-epops. Accessed 22 August 2017.
European Commission. "The Birds Directive." *European Commission: Environment*. http://ec.europa.eu/environment/nature/legislation/birdsdirective/index_en.htm. Accessed 16 May 2011.
———. "Environment: Commission Requests Malta to Comply with Court Ruling on Bird Hunting." *European Commission Press Release Database*. 28 October 2010. http://europa.eu/rapid/pressReleasesAction.do?reference=IP/10/1409&format=HTML.
Greer, Kirsten. "Transnational Ecologies in Malta: How Can a Critical Historical Geography Approach Shed Light on the Current Spring Bird Hunting Issue in Malta?" *NiCHE: Network in Canadian History & Environment*. 18 June 2009. http://niche-canada.org/2009/06/18/transnational-ecologies-in-malta/.
"Jill Malusky and Henry McGhie (Head of Natural Sciences and Curator of Zoology)." YouTube video, 2:39, from a discussion on the Manchester Museum's bird collection, its origins and relationship to the Revealing Histories project,

posted by "mzfasmjc," 29 June 2007. http://www.youtube.com/watch?v
=Qre1UaM-roE.
McCarthy, Michael. "Conservationists Mobilise to Halt Mass Slaughter of Birds in Malta." *Independent*, 25 February 2008. http://www.independent.co.uk /environment/nature/conservationists-mobilise-to-halt-mass-slaughter-of-birds -in-malta-786763.html.
Morris, Lydia. "Rare Exotic Hoopoe Bird Spotted on Pen Llŷn in Gwynedd." *Daily Post*, 18 August 2015. http://www.dailypost.co.uk/news/north-wales-news /hoopoe-bird-spotted-gwynedd-rspb—9880329.
O'Connor, D. P. "The RUSI, Imperial Defence and the Expansion of Empire 1829–90." *Royal United Services Institute (RUSI)*. http://www.rusi.org/downloads /assets/OConnor,_The_Influence_of_the_RUSI.pdf. Accessed 7 June 2011.
Rappaport, Helen. "Christmas in the Crimea." *Dovegrey reader scribbles*, 22 December 2011. http://dovegreyreader.typepad.com/dovegreyreader_scribbles /2011/12/christmas-in-the-crimea-a-guest-blog-post-by-helen-rappaport.html.
RSPB (Royal Society for the Protection of Birds). "Bringing an End to the Maltese Killing Fields." Last modified 25 June 2010. https://ww2.rspb.org.uk/our-work /rspb-news/news/details.aspx?id=tcm:9-254874.
———. "Ospreys in the UK." https://ww2.rspb.org.uk/our-work/conservation /conservation-and-sustainability/safeguarding-species/case-studies/osprey /#r7i9bwgMvo8LIJ13.99. Accessed 17 July 2019.
Sears, Neil. "Michael Palin: 'Britons Should Stop Apologising for Their Colonial Past and Be Proud of Our Empire's Achievements.'" *Daily Mail* online, 2 October 2009.
TVM. "Birdlife Malta Loses Libel Case against FKNK." 1 February 2016. https://www .tvm.com.mt/en/news/birdlife-malta-loses-libel-case-against-fknk/.
Zammit, Peter. "Malta No Mecca for Migratory Birds." *Times of Malta*, 24 September 2010. .

# Index

Note: Page numbers in italics denote illustrative material.

Abderahman, Moulay, Sultan of Morocco, 133n104
acclimatization, 49
Adams, Andrew Leith, 44–62, 100–101; climate studies in India, 49–52; climate studies in Malta, 52–54, 127n137; climate studies in New Brunswick, 58–60; contributions to museums, 49, 58, 124n71, 127n153; geographic route, 42; on golden orioles, 63; and heterogeneity of military ornithologist experience, 101–3; on hoopoes, 41; labeling practices, 112n8; on Malta, 44–45; medical training, 47–48, 124n70; on ospreys, 81; studies on connectivity of Europe to Africa, 55–58; on value of ornithology, 45, 93; writings, 19–20, 44–45, 53, 122n35, 127n148
Adams, Bertha Grundy, 54
Adams, Francis (father of Andrew Adams), 47, 123n52
Adams, Francis (son of Andrew Adams), 126n115, 128n161
Albert, Prince Consort of England, 28
Aldershot, Hampshire: fieldwork in countryside of, 89–92; as imperial home station, 84–85, 86–87
Alexander, James, 35, 119n91
Allen, E. G., 4
Amazigh people, 133n107
Aristotle, 45–46
Arnold, David, 49
Arnold, Sergeant, 86
*Atlas of Zoogeography* (Bartholomew), 21, 116n95

authentication process, 38–39
avian imperial archive: as concept, 11–14; and zoogeographic mapping, 15–17, 19, 21

Bartholomew, J. G., 21, 116n95
Bartlett, William Henry, 70
Bayly, Christopher A., 130n31
Becher, E. F., 16, 73–74
Becker, Cynthia J., 133n107
Bengalis, 68
Bermuda, 82–83
Bird, James, 51
birds. *See* migrant bird life geographies
Blackburn, Jemima (née Wedderburn), 54, 126n114
Blackwood, Alicia, 30, 117n50
Blagden, Charles, 4, 30
Blakiston, John, 31, 118n54, 118n59
Blakiston, Lawrence, 34, 119n83
Blakiston, Thomas Wright, 26–40, 100; contributions to museums, 37–39; on egg identification, 20; geographic route, 24; on great bustards, 23; and heterogeneity of military ornithologist experience, 101–3; military training and career, 30–32, 40, 118n57; in Nova Scotia, 32, 59, 128n162; ornithological work during Crimean War, 32–36; relationship with Downs, 32, 118n66
Blakiston family, 118n51
Bland, Edward Loftus, 32, 118n65, 128n162
Blunt, Alison, 85
Boardman, George Augustus, 128n174

167

Bonhomme, Brian, 85
Boudin, Jean Christian, 122n41
Boulton, Charles A., 87
Bree, Charles, 55
British Association for the Advancement of Science (BAAS), 17, 30, 34, 114n60
*British Birds: Key List* (Irby), 10
Britishness: of Hampshire countryside, 89–92; in India, 31, 51, 67–68; and national bird discourse, 91–95, 101; vs. pothunters, 36, 76, 78
British Ornithological Union (BOU), 10, 16, 57–58, 84, 88, 92–93, 135n45, 135n48, 137n104
Brown, Robert, 21
Browne, Janet, 15

Camerini, Jane, 21
Cardwell Reforms, 122n35
Carte, William, 35, 37, 119n91
*Catalogue of the Collection of Mammalia and Birds in the Museum of the Army Medical Department at Fort Pitt, Chatham, A*, 48
Catesby, Mark, 4, 110n19
cholera, 57, 126n110
Clarke, W. Eagle, 116n95
class: of military officers, 53, 72, 75; and pothunting, 36, 76, 119n100
classification systems, 18, 34, 92–94, 112n8, 114n60, 137n105
climate. *See* tropicality
Coleridge, Samuel Taylor, 70
Colley, Linda, 53, 109n11, 133n109
colonialism. *See* imperial place-making
Congreve, William, 120n117
Connors, Reverend Mr., 30
Convention on Migratory Species, 65–66, 109n3
Convention on the Protection and the Promotion of the Diversity of Cultural Expressions, 1
countryside, English, 89–92
Coventry, John, 75
Cowper, William, 95

Crimean Tartars, 36, 119nn103–4
Crimean War (1853–56): and avian scientific trophies of war, 37–39; and heroic imaginary, 27–30; ornithological work during, 32–36
Crimean War Memorial (Royal Artillery Barracks), 38, 120n118
Cundall, Joseph, 28
Cushman, Gregory, 111n38

Dartford warblers, 94–95
Darwin, Charles, 15, 31
Darwin, John, 56
Davies, Thomas, 18, 115n73
Davy, John, 52, 125n97
Dawson, Graham, 27, 33, 68
Denison, William, 53, 136n69
Disraeli, Benjamin, 125n94
domesticity, 30
Dowling, Robyn, 85
Downes, Henry, 29
Downs, Andrew, 32, 118n66
Dresser, Arthur, 128n163
Dresser, Henry, 59, 112n19, 128n163, 128n174
Drinkwater, John, 70
Driver, Felix, 113n32
Drummond Hay, Maurice, 41, 44, 78, 126n129, 133n104, 135n47
Dual Control (1879–82), 56
Duncan, James, 46

eagles, 70, 73, 74
East India Company, 31, 49–50, 124n66
Edwards, George, 4, 16, 129n3
eggs, study of, 20
Egypt, 43, 56, 57–58, 127n153
empire. *See* imperial place-making; transimperialism
Englishness. *See* Britishness
ethical ornithology and preservationism, 3, 75–76, 77–78, 85, 91–92, 101, 110n18, 137n83
Evans, Emyr, 12
evolutionary theories, 15

168 *Index*

*Fabulous Histories* (Trimmer), 85
Favier, M. F., 44, 63, 79
Febvre, Lucien, 9
Feilden, Henry Wemyss, 81, 88, 135n53, 138n113
Fenton, Roger, 28
Ferguson, Moira, 85
Fort Pitt Museum of Natural History at Chatham, 49
Foucault, Michel, 67
Free Church of Scotland, 53
French military ornithologists, 35–36, 58, 78–79, 120n107
French territorial interests, 56, 127n153

Gamble, K., 38
Garratt, G. T., 76–77
garrison libraries, 30, 53, 75, 87, 125n101
gender. *See* masculinity
George III, King of England, 120n117
Gibraltar: citizenship, 70, 131n55; ethical ornithology in, 75–76; field sites around, 76–79; golden orioles in, 63, 129n3; great bustards in, 25; hoopoes in, 42; in imperial discourses, 68–71, 130n36; ospreys in, 82; strategic location of, 2, 66, 67, 68, 130n22
Gibraltar Garrison Library, 75
Gibraltarians, 71
Gilbard, Major, 70
*Gleanings on the Overland Route* (Bartlett), 70
Godwin-Austen, Henry Haversham, 131n67, 135n45
golden oriole (*Oriolus oriolus*), 63–66, 65, 129n3
Gordon, Arthur, 128n174
Gould, John, 38–39, 55, 92–93
Graham, Captain, 38
Grant, James Augustus, 48
Grant, Ulysses S., 70
Graves, Thomas, 55
Gray, John Edward, 31

great bustard (*Otis tarda*), 23–26, 24
Green, Martin Burgess, 28
Griffin, Lieutenant, 38
Grimshaw, Percy H., 116n95
*Guide to the Gallery of Birds in the Department of Zoology, British Museum*, 41

Hadfield, Henry, 94, 138n111
Halifax, Nova Scotia, 32, 59, 118n63
Hampshire. *See* Aldershot, Hampshire
*Hand-book to the Birds of Great Britain* (Sharpe), 92
*Handbook to the Mediterranean* (Playfair), 56
Hannah, Matt, 18
Haraway, Donna, 13
Harding, Colonel, 37, 120n110
Harley, J. B., 113n27
Hawker, Peter, 31
Hawker, William Henry, 37
Hay, Arthur, 28, 34
health tourism, 53
Henry, Captain, 37
heroic figure. *See* military-scientific hero
Hindus, 71
Hippocrates, 47
*Historical Atlas of Breeding Birds in Britain and Ireland 1875–1900, The*, 122n23
*History of British Birds, A* (Yarrell), 34, 110n19
*History of British Birds, Indigenous and Migratory, in Five Volumes, A* (MacGillivray), 48
*History of the British Possessions in the Mediterranean* (Martin), 70
*History of the Siege of Gibraltar* (Drinkwater), 70
Hooker, Joseph Dalton, 31
Hooker, William Jackson, 31
hoopoe (*Upupa epops*), 41–44, 43
Howe, Kathleen Stewart, 69
Howell, Philip, 2
Howlett, Robert, 28

Index   169

Hudson, W. H., 16, 84, 134n16
Humboldt, Alexander von, 18, 19, 50, 124n74
Hume, Edgar, 4
Hunt, James, 122n41
Huxley, Thomas Henry, 31

*Ibis* (journal), 16, 57, 57
identity. *See* Britishness; class; masculinity; military-scientific hero; race
*Illustrated London News*, 28, 70–71, 87
*Illustrations of British Ornithology* (Selby), 75
imperial place-making: and avian imperial archive, 11–14; and avian scientific trophies of war, 37–39; connection to English "homeland," 34, 51, 52, 85, 89–92; in Gibraltar, 68–71, 130n36; and imperial expansion, 12, 49, 56–58, 84, 99, 124n66; and national heroic imaginary, 27–30; and territorial military presence, 66–67, 68–69, 79; and zoogeographic regional mapping, 15–17, 16, 19, 21, 22, 98. *See also* transimperialism
India, 2, 31, 49–52, 67–68, 125n90
*Instructions to Young Sportsmen in All That Relates to Guns and Shooting* (Hawker), 31
*Introduction to the Birds of Great Britain* (Gould), 92
Irby, Charles Leonard, 129n18
Irby, Edward Methuen, 129n18
Irby, Frederick Paul, 129n18
Irby, Leonard Howard Lloyd, 66–80, 101; authentication of Blakiston's specimens, 38; biographical sketch, 129n18; in BOU, 135n45; contributions to county bird lists, 94; contributions to museums, 10, 17, 88, 94–95; on ethical ornithology, 75–76, 77–78, 101; exchanges with French ornithologists, 36; exchanges with Russian ornithologists, 35; on Favier, 79; geographic route, 64; on Gibraltar, 66; on golden orioles, 63; on great bustards, 23–25; and heterogeneity of military ornithologist experience, 101–3; on hoopoes, 42, 44; as military-scientific hero, 71–72; on ospreys, 82; on racial identity, 68; relationship with Reid, 87–88; writings, 20, 66, 129n18, 130n21

Jacobs, Nancy, 99
Jameson, Robert, 48
Jardine, William, 19, 55, 112n8, 118n63, 122n23
Jews, 70, 71
journals, travel, 19–20

Kant, Immanuel, 118n54
Kelham, H. R., 88, 135n45
Kelsall, J. E., 93–94
Kennedy, Alexander W. M. Clark, 135n45
Kennedy, Dane, 51, 53
Kingsley, Charles, 89
Knepper, Paul, 54

Laffan, Robert Michael, 86
Lambert, David, 5–6, 71
Latour, Bruno, 113n32, 136n76
Lefroy, John, 120n117
Lester, Alan, 5–6, 111n42, 111n44
L'Estrange, Paget Walter, 10, 135n45
levanter winds, 69
libraries, 30, 53, 75, 87, 125n101
life geographies, as analytical approach, 5–7, 13–14, 101–3
Lilford, Lord, 65, 89
Linnaean classification system, 18
Livingstone, David: connection to Southampton, 117n32, 136n56; as scientific hero figure, 28, 34
Livingstone, David N.: on climate, 46–47, 49, 124n67; on field vs. museum naturalists, 114n64; on zoogeography, 110n27

Lloyd, John Hays, 135n45
Lyell, Charles, 31, 56

MacDonald, Helen, 5
MacGillivray, William, 48, 123n59, 124n74
MacLeod, Roy, 4-5
Maliseet people, 59-60
Malta: British possession, 126n128; cholera in, 126n110; geographic connection with North Africa, 55-56; golden orioles in, 63; hoopoes in, 41, 44; ospreys in, 81; relationship with EU, 1; as semitropical, 45, 52-54; strategic location of, 2, 52, 125n94
Malta Garrison Library, 53, 125n101
*Malta Times* (newspaper), 53-54
Maltese people, 54
*Manual of British Ornithology, A* (MacGillivray), 48
*Manual of Zoology* (Newton), 21
mapping, zoogeographic, 15-17, *16*, 19, 21, 22, *98*
Marischal College, University of Aberdeen, 47-48
Martin, Robert Montgomery, 55, 70
masculinity: of military-scientific hero, 28-29, 71-72, 73, 77, 101; temperate, 28-29, 30, 45-46, 50-51, 53, 75, 89-90, 100-101
Massey, Doreen, 111n40, 134n22
Matless, David, 85, 136n65
McCarthy, Michael, 1, 2, 109n6
McGhie, Henry, 112n19
McGrigor, James, 48-49, 123n61
Mearns, Barbara and Richard, 4, 110n22, 138n114
medical tradition: in British Army, 47-48; on tropicality, 45-46, 49-53
migrant bird life geographies: as analytical approach, 6-7, 13-14; golden oriole (*Oriolus oriolus*), 63-66, *65*, 129n3; great bustard (*Otis tarda*), 23-26, *24*; hoopoe (*Upupa epops*), 41-44, *43*; osprey (*Pandion haliaetus*), 81-84, *83*, 134n11
Mi'kmaq people, 59-60
military ornithology and ornithologists: development of methods, 17-21; emergence of, 16-17; heterogeneity of experiences, 101-3; as phenomenon, overview, 3-5, 12, 14-15. *See also* Adams, Andrew Leith; Blakiston, Thomas Wright; Irby, Leonard Howard Lloyd; Reid, Philip Savile Grey
military reforms, 29-30, 39, 46, 86-87, 122n35
military-scientific hero: and avian scientific trophies of war, 37-39; Blakiston's representation of, 33-35; as concept, 26-27; as exemplar, 28-30, 100; Irby's representation of, 71-72; masculinity of, 28-29, 71-72, 73, 77, 101; and morale during Crimean War, 27-28
monogenism, 55
Montagu, George, 94, 138n114
Montagu harriers, 94
Moors, 70, 71, 79
morality: ethical ornithology and preservationism, 3, 75-76, 77-78, 85, 91-92, 101, 110n18, 137n83; of temperate masculinity, 28-29, 45-46, 50-51, 53, 75, 89-90, 100-101
More, Alexander Goodman, 92, 137n91
Morocco, 77-78, 79
Museum of the Royal Society of Dublin, 35, 37

Napier, Edward, 70, 76, 131n48
national birds, 91-95, 101
national hero. *See* military-scientific hero
national identity. *See* Britishness
Natural History Museum (South Kensington), 10, 65, 94-95, 111n1
*Natural History of Carolina, Florida and the Bahama Islands* (Catesby), 110n19

*Natural History of Selborne, The* (White), 75, 90–91, 110n19, 136n73
naturalists, field vs. museum, 18, 48, 114n64
*Nature* (journal), 71
Neilson, Andrew, 36, 119–20n105
nests, study of, 20, 65, 94–95
New Brunswick, 32, 58–60, 128n158
Newman, Edward, 39
Newton, Alfred, 21, 38, 58, 92, 93, 121n120, 137n83
Newton, Edward, 135n47
Newton, William Samuel, 121n120
nidification, 20
North Africa, 15, 17, 21–22, 44, 55–58, 77–79
*Notes of a Naturalist in the Nile Valley and Malta* (Adams), 44–45
Notman, William, 138n1
Nova Scotia, 32, 59, 128n162

Ó Cadhla, Stiofán, 67
Ogborn, Miles, 111n44
oology, 20
ornithology. *See* military ornithology and ornithologists
*Ornithology of the Straits of Gibraltar, The* (Irby), 66, 129n18
osprey (*Pandion haliaetus*), 81–84, 83, 134n11
Ottoman Empire, 56
*Our Earth and Its Story: A Popular Treatise on Physical Geography* (Brown), 21

Paasi, Anssi, 66–67
Palliser, John, 40
Peers, Douglas, 53
Perry, Charles, 135n39
Perry, W. Edward, 135n39
Phillips, Richard, 77
photography, 20, 28, 73
Pindar, John, 71, 75, 131n59
place, as concept, 6–7. *See also* imperial place-making
Playfair, Robert Lambert, 56

polygenism, 55
Portlock, Joseph Ellison, 29–30, 117n40
pothunting, 1, 2, 36, 54, 76, 78, 91, 119n100
Pratt, Mary Louise, 34, 76
preservationism and ethical ornithology, 3, 75–76, 77–78, 85, 91–92, 101, 110n18, 137n83

race: and degeneration, 59–60, 67–68; diversity in Gibraltar, 70–71; and pothunting, 54; whiteness, 50, 54, 59
Raffles, Hugh, 21, 136n76
Ranken, George, 23, 34
Rawlinson, H. C., 29
Reeve, William, 88
Reform Act (1832), 118n54
Reid, Philip Savile Grey, 84–96, 101; contributions to county bird lists, 93–94; contributions to museums, 10, 95; on ethical ornithology, 91–92; fieldwork in English countryside, 90–91, 94, 136n69; geographic route, 82; and heterogeneity of military ornithologist experience, 101–3; metropolitan scientific networks, 87–89; military training, 134-35n39; on ospreys, 82–83; on pothunting, 76; on robin redbreast, 93; on Stark, 132n100; studies on eagles, 74; studies on levanter winds, 69
Reid, William, 53, 125n102
Rhodes, William, 138n1
Rich, Adrienne, 14
Richards, Thomas, 12, 112n12
Richardson, David Lester, 67, 86
Richardson, John, 31, 49, 123n63
Ritter, Carl, 32, 118n60
Ritvo, Harriet, 36
robin redbreast, 34, 93, 95
Rose, Sonya, 46
Royal Artillery Institution (RAI), 37–38, 120n117
Royal Geographical Society, 17, 19, 30, 116n95

172   Index

Royal Military Academy, 31–32, 89–90
Royal Society, 17, 30
Royal Society for the Protection of Birds, 65
Royal United Service Institution (former United Service Institution), 29
Russian military ornithologists, 35

Sabine, Edward, 4, 30, 40
St. John, Oliver Beauchamp Coventry, 135n45
Sanchez, M. G., 71
Saunders, Howard, 93
savage nature, 59–60, 67–68
Schembri, Signor, 56
Scicluna, Sandra, 54
Sclater, Philip L., 15–16, 17, 21, 114n46
Sclater, William, 21
Seacole, Mary, 23
Second Anglo-Sikh War, 124n66
Selby, Prideaux John, 75
semitropicality, 52–54, 100–101
Sharpe, Richard Bowdler, 10, 92, 95, 112n7
Shelley, George Ernest, 10, 58, 135n45
Siege of Sebastopol (1855–56), 33
Sikh Empire, 124n66
sirocco winds, 52
sketching, 20, 73
*Soldier's Pocket-Book for Field Service* (Wolseley), 73
Souper, C. E., 38
Southampton, 135–36n56
South Kensington Natural History Museum, 10, 65, 94–95, 111n1
Spaniards, 70, 76
Spanish territorial interests, 69, 76–77
Spratt, Captain, 56
Stagl, Justin, 113n38
Stanley, Peter, 49–50
Stark, Arthur Cowell, 132n100
Strickland, Hugh Edwin, 55, 126n129
Strickland code, 114n60
Suez Canal, 56, 66, 67, 86, 125n94
Swainson, William, 49, 123n63

Tangier, 78–79, 133n109
Tartar people, 36, 119nn103–4
taxidermy, 18, 32
taxonomies, 18, 34, 92–94, 112n8, 114n60, 137n105
Taylor, George Cavendish, 36, 43, 63
Tebbutt, M., 136n65
temperance movements, 29, 48, 53
temperate, the: and masculinity, 28–29, 30, 45–46, 50–51, 53, 75, 89–90, 100–101; vs. "savage" nature, 59–60, 67–68; and tropicality, 45–46, 49–52
territoriality. *See* imperial place-making; transimperialism
Thackeray, G., 64
Thackeray, William Makepeace, 53, 70
time-space compression, 130n29
transimperialism: of Aldershot military network, 84–85, 86–87; and English "homeland," 34, 51, 52, 85, 89–92; Europe's geographic connection with North Africa, 21–22, 55–58, 77; of metropolitan scientific network, 87–88; and notions of tropicality and the temperate, 45–46, 49–54; strategic empire routes, 2, 17, 27, 33, 52, 66, 67, 68, 99, 130n22. *See also* imperial place-making
travel journals, 19–20
Trimmer, Sarah, 85
Tristram, Canon, 112n8
Tristram, Henry, 58, 127n154
tropicality: as concept, 45–47; in India, 49–52; vs. New Brunswick climate, 58–60; semitropicality, 52–54, 100–101
Tunis, 56–57, 127nn144–45
Turkey, 43
*Twelve Years of Military Adventures in Three Quarters of the Globe* (Blakiston), 31
*Twenty Years in Retirement* (Blakiston), 31
type localities, 19, 99
type specimens, 18, 115n72

United Service Institution (later Royal United Service Institution), 29
*United Service Magazine*, 29, 75

Verner, Willoughby, 10, 73, 74, 77, 130n21, 132n72
Victoria, Queen of England, 28, 30, 87

Wallace, Alfred Russel, 15, 19, 53, 92, 125n107
Wallerstein, Immanuel, 109n8
*Wanderings of a Naturalist in India* (Adams), 53
Warren, Emily Mary Bibbens, 138n125
Watkins, C. W., 42, 77
Watson, H. C., 92
Wedderburn, H., 44
Wedderburn, John Walter, 32, 118n63
Welch, F. H., 127n148
White, Gilbert, 34, 75, 90–91, 110n19, 136n73
White, John, 82, 110n19
Whitely, Henry, 38, 120–21n119
whiteness, 50, 54, 59. *See also* race
wilderness, 28, 59–60
Wilson, Robert, 5, 48, 123n55
winds, 52, 69
Wisely, George, 48
Wolseley, Garnet, 73
women, imperial contributions, 53–54, 87
Woodward, Rachel, 134n26
Wright, Charles Augustus, 44, 54, 55–56, 81, 125n107, 126n114, 126n135, 127n144

Yarrell, William, 34, 110n19

zoogeography: as field, 7, 15; regional mapping, 15–17, 16, 19, 21, 22, 98
Zoological Gardens (London), 37, 64, 87–88
Zoological Society of London, 16, 78
*Zoologist* (journal), 26, 33, 39, 90

www.ingramcontent.com/pod-product-compliance
Lightning Source LLC
Chambersburg PA
CBHW030655230426
43665CB00011B/1100